Universitext

Universitext

Universitext is a series of textbooks that presents material from a wide variety of mathematical disciplines at master's level and beyond. The books, often well class-tested by their author, may have an informal, personal even experimental approach to their subject matter. Some of the most successful and established books in the series have evolved through several editions, always following the evolution of teaching curricula, to very polished texts.

Thus as research topics trickle down into graduate-level teaching, first textbooks written for new, cutting-edge courses may make their way into *Universitext*.

For further volumes:
http://www.springer.com/series/223

Jan Vrbik • Paul Vrbik

Informal Introduction to Stochastic Processes with Maple

 Springer

Jan Vrbik
Department of Mathematics
Brock University
St Catharines
Ontario, Canada

Paul Vrbik
Department of Computer Science
The University of Western Ontario
London, Ontario, Canada

Additional material to this book can be downloaded from http://extras.springer.com

ISSN 0172-5939 ISSN 2191-6675 (electronic)
ISBN 978-1-4614-4056-7 ISBN 978-1-4614-4057-4 (eBook)
DOI 10.1007/978-1-4614-4057-4
Springer New York Heidelberg Dordrecht London

Library of Congress Control Number: 2012950415

Mathematics Subject Classification (2010): 60-01, 60-04, 60J10, 60J28, 60J65, 60J80, 62M10

Printed on acid-free paper

Springer is part of Springer Science+Business Media (www.springer.com)

Preface

This book represents a consolidation of lectures given by one of the authors over a 25-year period. It is intended for students who wish to apply stochastic processes to their own field.

Our goal is to use an informal, simple, and accessible style to introduce all basic types of these processes. Most of our examples are supplemented by MAPLE programs (MAPLE's syntax resembles pseudocode and can easily be adapted to other systems), including Monte Carlo simulations of each particular process. This enables the reader to better relate to the corresponding theoretical issues.

The classic texts in this subject area are too dated to utilize modern computer-algebra systems to, for instance, manipulate generating functions or build numerical rather than analytic solutions. Consequently, these techniques have been ignored historically because they were totally impractical when working strictly by hand. Since computers are now pervasive, fully integrating their usage into our text is a major contribution of the book. In fact, this, combined with our belief that overemphasizing mathematical details makes the material inaccessible to students, was our motivation.

In our writing we strive to satisfy three simple criteria: readability, accessibility, and brevity. To be *readable* we write informally, encouraging the reader to ask meaningful questions first and then systematically resolve them, one by one. In this way, our narrative should be confluent and fluid, so that reading the book cover to cover is not only possible but, hopefully, enjoyable.

To be *accessible*, we use ample examples throughout, accompanying each new notion or result with a specific illustration of some real-world application. Many of these are MAPLE simulations of the corresponding process, illustrating its main features.

We also delegate to MAPLE the derivation of some formulas, demonstrating its usefulness for algebraic manipulation. At the same time, we try to be as rigorous as possible, formally proving practically all our assertions. We usually do so verbally, thereby avoiding complex mathematical notation whenever possible.

Similarly, we have been careful not to assume much mathematical knowledge—whenever a new technique or concept is needed, an introduction to the corresponding mathematical background is provided.

Finally, *brevity* was a natural consequence of our goal to be concise. It was important to us to provide a framework for designing a two-semester course.[1] A book of fewer than 300 pages certainly fits this criterion.

We would like to acknowledge and thank Dr. Rob Corless for his help and encouragement, as well as Brandon Clarke for pointing out many of our grammatical errors.

Ontario, Canada   Jan Vrbik
Ontario, Canada   Paul Vrbik

---

[1] Lecturers designing one- or two-semester courses should be aware that Chaps. 4, 5, 10, and 11 are self-contained, whereas Chaps. 2–3 and Chaps. 6–9 constitute a natural sequence.

Contents

CHAPTER 1
Introduction

A STOCHASTIC (a fancy word for "random") PROCESS is a *collection* (often infinite, at least in principle) *of random variables*, labeled by a parameter (say) t, which represents time. The random variables are usually denoted by $X(t)$ when t has a continuous scale of real values and X_t when t is restricted to integers (e.g., day 1, day 2).

Example 1.1 (Trivial Stochastic Process). A random independent sample from a specific distribution of infinite size, that is, X_1, X_2, X_3, ..., is the simplest example of a stochastic process. $\diamond$

A more typical stochastic process will have individual random variables *correlated* with one another.

Stochastic processes are of four rather distinct categories, depending on whether the values of X_t and of t are of a discrete or continuous type. The individual categories are as follows.

BOTH X_t AND t SCALES ARE DISCRETE

Example 1.2 (Bernoulli Process). Flipping a coin repeatedly (and indefinitely). In this case, X_1, X_2, X_3, ... are the individual outcomes (the STATE SPACE consists of -1 and 1, to be interpreted as losing or winning a dollar). $\diamond$

Example 1.3 (Cumulative Bernoulli Process). Consider the same Bernoulli process as in Example 1.2, where Y_1, Y_2, Y_3, ... now represent the CUMULATIVE SUM of money won so far (i.e., $Y_1 = X_1$, $Y_2 = X_1 + X_2$, $Y_3 = X_1 + X_2 + X_3$, ...). This time the Y values are correlated (the state space consists of all integers). $\diamond$

J. Vrbik and P. Vrbik, *Informal Introduction to Stochastic Processes with Maple*, Universitext, DOI 10.1007/978-1-4614-4057-4_1,
© Springer Science+Business Media, LLC 2013

Example 1.4 (Markov Chains). These will be studied extensively during the first part of the book (the sample space consists of a handful of integers for *finite* Markov chains and of *all* integers for *infinite* Markov chains).   ◇

X_t DISCRETE, t CONTINUOUS

Example 1.5 (Poisson Process). The number of people who have entered a library from time zero until time t. $X(t)$ will have a Poisson distribution with a mean of $\lambda \cdot t$ (λ being the average arrival rate), but the X are not independent (Fig. 6.1 for a graphical representation of one possible REALIZATION of such a process – the sample space consists of all nonnegative integers).   ◇

Example 1.6 (Queuing Process). People not only enter but also leave a library (this is an example of an infinite-server queue; to fully describe the process, we need also the distribution of the time a visitor spends in the library). There are also queues with one server, two servers, etc., with all sorts of interesting variations.   ◇

BOTH X_t AND t CONTINUOUS

Example 1.7 (Brownian Motion). Also called DIFFUSION – a tiny particle suspended in a liquid undergoes an irregular motion due to being struck by the liquid's molecules. We will study this in one dimension only, investigating issues such as, for example, the probability the particle will (ever) come back to the point from which it started.   ◇

X_t CONTINUOUS, t DISCRETE

Example 1.8 (Time Series). Monthly fluctuations in the inflation rate, daily fluctuations in the stock market, and yearly fluctuations in the Gross National Product fall into the category of time series. One can investigate trends (systematic and seasonal) and design/test various models for the remaining (purely random) component (e.g., Markov, Yule). An important issue is that of estimating the model's parameters.   ◇

In this book we investigate at least one type of each of the four categories, namely:

1. Finite Markov chains, branching processes, and the renewal process (Chaps. 1–4);
2. Poisson process, birth and death processes, and the continuous-time Markov chain (Chaps. 5–8);
3. Brownian motion (Chap. 9);
4. Autoregressive models (Chap. 10).

Solving such processes (for any *finite* selection of times t_1, t_2, ..., t_N) requires computing the distribution of each individual $X(t)$, as well as the *bivariate* distribution of any $X(t_1)$, $X(t_2)$ pair, trivariate distribution of any $X(t_1)$, $X(t_2)$, $X(t_3)$ triplet, and so on. As the multivariate cases are usually simple extensions of the univariate one, the univariate distributions of a single $X(t)$ will be the most difficult to compute.

Yet, depending on the type of process being investigated, the mathematical techniques required are surprisingly distinct. We require:

- All aspects of matrix algebra and the basic theory of difference equations to handle finite Markov chains;
- A good understanding of function composition and the concept of a sequence-generating function to deal with branching processes and the renewal theory;
- A basic (at least conceptually) knowledge of partial differential equations (for Chaps. 5–7);
- Familiarity with eigenvalues of a square matrix to learn how to compute a specific function of any such matrix (for Chap. 8); and, finally,
- Calculus (Chaps. 5, 9–10) and complex number manipulation (Chaps. 2, 8, and 10).

In an effort to make the book self-contained, we provide a brief overview of each of these mathematical tools in the chapter appendices.

We conclude this section with two definitions:

Definition 1.1 (Stationary). A process is STATIONARY when all the X_t have the same distribution, and also: for each τ, all the $(X_t, X_{t+\tau})$ pairs have the same bivariate distribution, similarly for triplets, etc.

Example 1.9. Our queueing process can be expected to become stationary (at least in the $t \to \infty$ limit, i.e., asymptotically), but the cumulative-sum process is nonstationary. $\diamond$

Definition 1.2 (Markovian property). A process is MARKOVIAN when

$$\Pr\left(X_{i+1} < x \mid X_i = x_i, X_{i-1} = x_{i-1}, \ldots, X_0 = x_0\right)$$
$$= \Pr\left(X_{i+1} < x \mid X_i = x_i\right),$$

or, more generally, to compute the probability of an event in the future, given a knowledge of the past and present, one can discard information about the past without affecting the answer. This does *not* imply X_{i+1} is independent of, for example, X_{i-1}, X_{i-2}.

Example 1.10. The stock market is most likely non-Markovian (trends), whereas the cumulative-sum process has a Markovian property. $\diamond$

The main objective in solving a specific stochastic-process model is to find the joint distribution of the process's values for any *finite* selection of the t indices. The most basic and important of these is the UNIVARIATE distribution of X_t, for any value of t, from which the multivariate distribution of several X_t (usually) easily follows.

CHAPTER 2
Finite Markov Chains

FINITE MARKOV CHAINS are processes with finitely many (typically only a few) states on a nominal scale (with arbitrary labels). Time runs in discrete steps, such as day 1, day 2, ..., and only the most recent state of the process affects its future development (the Markovian property). Our first objective is to compute the probability of being in a certain state after a specific number of steps. This is followed by investigating the process's *long-run* behavior.

2.1 A FEW EXAMPLES

To introduce the idea of a Markov chain, we start with a few examples.

Example 2.1. Suppose that weather at a certain location can be sunny, cloudy, or rainy (for simplicity, we assume it changes only on a daily basis). These are called the STATES of the corresponding process.

The simplest model assumes the type of weather for the next day is chosen randomly from a distribution such as

Type	S	C	R
Pr	$\frac{1}{2}$	$\frac{1}{3}$	$\frac{1}{6}$

(which corresponds to rolling a biased die), *independently* of today's (and past) conditions (in Chap. 1, we called this a *trivial* stochastic process).

Weather has a tendency to resist change, for instance, sunny → sunny is more likely than sunny → rainy (incidentally, going from X_n to X_{n+1} is called a TRANSITION). Thus, we can improve the model by letting the distribution depend on the current state. We would like to organize the corresponding information in the following TRANSITION PROBABILITY MATRIX (TPM):

J. Vrbik and P. Vrbik, *Informal Introduction to Stochastic Processes with Maple*, Universitext, DOI 10.1007/978-1-4614-4057-4_2,
© Springer Science+Business Media, LLC 2013

	S	C	R
S	0.6	0.3	0.1
C	0.4	0.5	0.1
R	0.3	0.4	0.3

where the rows correspond to today's weather and the columns to the type of weather expected tomorrow (each row must consist of a complete distribution; thus all the numbers must be nonnegative and sum to 1). Because tomorrow's value is not *directly* related to yesterday's (or earlier) value, the process is *Markovian*.

There are several issues to investigate, for example:

1. If today is sunny, how do we compute the probability of its being rainy *two* days from now (three days from now, etc.)?
2. In the *long run*, what will be the proportion of sunny days?
3. How can we improve the model to make the probabilities depend on today's *and* yesterday's weather?

To generate a possible realization of the process (starting with sunny weather) using MAPLE, we type

```
> with(LinearAlgebra): with(plots): with(Statistics):
```

{Henceforth we will assume these packages are loaded and will not explicitly call them (see "Library Commands" in Chap. 13).}

$$> \mathbb{P}_1 := \begin{bmatrix} 0.6 & 0.3 & 0.1 \\ 0.4 & 0.5 & 0.1 \\ 0.3 & 0.4 & .3 \end{bmatrix} :$$

```
> (j, res) := (1, 1) :
> for i from 1 to 25 do
      j := Sample (ProbabilityTable (convert (ℙ₁[j], list)) , 1)₁ ;
      j := trunc(j);
      res := res, j;
    end do:
> subs (1 = S, 2 = C, 3 = R, [res]) ;
```

$$[S, S, C, R, S, S, S, C, C, C, R, S, S, R, R, R, C, S, S, S, S, C, S, S, S, S]$$

(The MAPLE worksheets can be downloaded from extras.springer.com.)

Example 2.2. Alice and Bob repeatedly bet \$1 on the flip of a coin. The potential states of this process are all integers, the INITIAL STATE (usually denoted X_0) may be taken as 0, and the TPM is now infinite, with each row looking like this:

$$\cdots \ 0 \ 0 \ \tfrac{1}{2} \ 0 \ \tfrac{1}{2} \ 0 \ 0 \ \cdots$$

This is an example of a so-called INFINITE MARKOV CHAIN. For the time being, we would like to investigate FINITE MARKOV CHAINS (FMCs) only, so we modify this example assuming each player has only \$2 to play with:

	-2	-1	0	1	2
-2	1	0	0	0	0
-1	$\tfrac{1}{2}$	0	$\tfrac{1}{2}$	0	0
0	0	$\tfrac{1}{2}$	0	$\tfrac{1}{2}$	0
1	0	0	$\tfrac{1}{2}$	0	$\tfrac{1}{2}$
2	0	0	0	0	1

The states are labeled by the amount of money Alice has won (or lost) so far. The two "end" states are called ABSORBING states. They represent the situation of one of the players running out of money; the game is over and the Markov chain is stuck in an absorbing state for good.

Now the potential *questions* are quite different:

1. What is the probability of Alice winning over Bob, especially when they start with different amounts or the coin is slightly biased?
2. How long will the game take (i.e., the distribution, expected value, and standard deviation of the number of transitions until one of the players goes broke)?

Again, we can simulate one possible outcome of playing such a game using MAPLE:

$$> \mathbb{P}_2 := \begin{bmatrix} 1 & 0 & 0 & 0 & 0 \\ \tfrac{1}{2} & 0 & \tfrac{1}{2} & 0 & 0 \\ 0 & \tfrac{1}{2} & 0 & \tfrac{1}{2} & 0 \\ 0 & 0 & \tfrac{1}{2} & 0 & \tfrac{1}{2} \\ 0 & 0 & 0 & 0 & 1 \end{bmatrix} :$$

```
> (j, res) := (3, 3) :
> for i from 1 while (j > 1 and  j < 5) do
      j := Sample (ProbabilityTable (convert (P₂[j], list)) , 1)₁ ;
      j := trunc(j);
      res := res, j :
   end do:
```

$$> subs(1 = -2, 2 = -1, 3 = 0, 4 = 1, 5 = 2, [res]);$$

$$[0, -1, 0, 1, 0, 1, 2]$$

(Note Alice won six rounds.)

Example 2.3. Assume there is a mouse in a maze consisting of six compartments, as follows:

```
┌──────┬───┰───┐
│      │   ┃   │
│  1   │ 2 │ 3 │
│      ┝━┿━┥┏━┥
├──┿━┥─┘   ┗─┤
│      │   │   │
│  4   │ 5 │ 6 │
│      ┃   ┃   │
└──────┸───┸───┘
```

Here we define a transition as happening whenever the mouse changes compartments. The TPM is (assuming the mouse chooses one of the available exits perfectly randomly)

	1	2	3	4	5	6
1	0	0	0	1	0	0
2	0	0	$\frac{1}{2}$	0	$\frac{1}{2}$	0
3	0	1	0	0	0	0
4	$\frac{1}{2}$	0	0	0	$\frac{1}{2}$	0
5	0	$\frac{1}{3}$	0	$\frac{1}{3}$	0	$\frac{1}{3}$
6	0	0	0	0	1	0

Note this example is what will be called PERIODIC (we can return to the same state only in an *even* number of transitions).

A possible REALIZATION of the process may then look like this (taking 1 as the initial state):

$$> \mathbb{P}_3 := \begin{bmatrix} 0 & 0 & 0 & 1 & 0 & 0 \\ 0 & 0 & \frac{1}{2} & 0 & \frac{1}{2} & 0 \\ 0 & 1 & 0 & 0 & 0 & 0 \\ \frac{1}{2} & 0 & 0 & 0 & \frac{1}{2} & 0 \\ 0 & \frac{1}{3} & 0 & \frac{1}{3} & 0 & \frac{1}{3} \\ 0 & 0 & 0 & 0 & 1 & 0 \end{bmatrix} :$$

```
> (j, res) := (1, 1) :
> for i from 1 to 30 do
      j := Sample (ProbabilityTable (convert (ℙ₃[j], list)), 1)₁ ;
      res := trunc(j);
      res := res, j;
  end do:
> res;
```

$$1, 4, 1, 4, 1, 4, 5, 6, 5, 2, 3, 2, 3, 2, 5, 4, 1, 4, 1, 4, 1, 4, 1, 4, 5, 4, 1, 4, 5, 6, 5, 2$$

One of the issues here is finding the so called FIXED VECTOR (the relative frequency of each state in a long run), which we discuss in Sect. 2.5.

We modify this example by opening Compartment 6 to the outside world (letting the mouse escape, when it chooses that exit). This would then add a new "Outside" state to the TPM, a state that would be absorbing (the mouse does not return). We could then investigate the probability of the mouse's finding this exit eventually (this will turn out to be 1) and how many transitions it will take to escape (i.e., its distribution and the corresponding mean and standard deviation). ◇

Example 2.4. When repeatedly tossing a coin, we may get something like this:

$$\text{HTHHHTTHTH}\ldots.$$

Suppose we want to investigate the patterns of two consecutive outcomes. Here, the first such pattern is HT, followed by TH followed by HH, etc. The corresponding TPM is

	HH	HT	TH	TT
HH	$\frac{1}{2}$	$\frac{1}{2}$	0	0
HT	0	0	$\frac{1}{2}$	$\frac{1}{2}$
TH	$\frac{1}{2}$	$\frac{1}{2}$	0	0
TT	0	0	$\frac{1}{2}$	$\frac{1}{2}$

This will enable us to study questions such as the following ones:

1. What is the probability of generating TT *before* HT? (Both patterns will have to be made absorbing.)
2. How long would such a game take (i.e., what is the expected value and standard deviation of the number of flips needed)?

The novelty of this example is the initial setup: here, the very first state will itself be generated by two flips of the coin, so instead of starting in a

specific initial state, we are randomly selecting it from the following INITIAL
DISTRIBUTION:

State	HH	HT	TH	TT
Pr	$\frac{1}{4}$	$\frac{1}{4}$	$\frac{1}{4}$	$\frac{1}{4}$

In Sect. 5.1, we will extend this to cover a general situation of generating
a pattern like HTTHH before THHT.   ◇

2.2 TRANSITION PROBABILITY MATRIX

It should be clear from these examples that all we need to describe a
Markov chain is a corresponding TPM (all of whose entries are ≥ 0 and whose
row sums are equal to 1 – such square matrices are called STOCHASTIC) and
the initial state (or distribution).

The one-step TPM is usually denoted by $\mathbb{P}$ and is defined by

$$\mathbb{P}_{ij} \equiv \Pr(X_{n+1} = j \mid X_n = i).$$

In general, these probabilities may depend on n (e.g., the weather patterns
may depend on the season, or the mouse may begin to learn its way through
the maze). For the Markov chains studied here we assume this does *not*
happen, and the process is thus HOMOGENEOUS IN TIME, that is,

$$\Pr(X_{n+1} = j \mid X_n = i) \equiv \Pr(X_1 = j \mid X_0 = i)$$

for all n.

TWO-STEP (THREE-STEP, ETC.) TRANSITION PROBABILITIES

Example 2.5. Suppose we have a three-state FMC, defined by the following
(general) TPM:

$$\mathbb{P} = \begin{bmatrix} p_{11} & p_{12} & p_{13} \\ p_{21} & p_{22} & p_{23} \\ p_{31} & p_{32} & p_{33} \end{bmatrix}.$$

Given we are in State 1 now, what is the probability that *two* transitions
later we will be in State 1? State 2? State 3?

Solution. We draw the corresponding probability tree

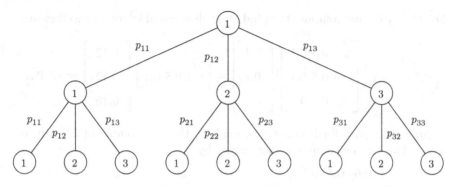

and apply the formula of total probability to find the answer $p_{11}p_{11} + p_{12}p_{21} + p_{13}p_{31}$, $p_{11}p_{12} + p_{12}p_{22} + p_{13}p_{32}$, etc. These can be recognized as the $(1,1)$, $(1,2)$, etc. elements of $\mathbb{P}^2$. $\qquad\qquad\Diamond$

One can show that in general the following proposition holds.

Proposition 2.1.

$$\Pr(X_n = j \mid X_0 = i) = (\mathbb{P}^n)_{ij}.$$

Proof. Proceeding by induction, we observe this is true for $n = 1$. Assuming that it is true for $n - 1$, we show it is true for n.

We know that $\Pr(A) = \sum_k \Pr(A \mid C_k)\Pr(C_k)$ whenever $\{C_k\}$ is a partition. This can be extended to $\Pr(A \mid B) = \sum_k \Pr(A \mid B \cap C_k)\Pr(C_k \mid B)$; simply replace the original A by $A \cap B$ and divide by $\Pr(B)$. Based on this generalized formula of total probability (note $X_{n-1} = k$, with all possible values of k, is a partition), we obtain

$$\Pr(X_n = j \mid X_0 = i)$$
$$= \sum_k \Pr(X_n = j \mid X_{n-1} = k \cap X_0 = i)\Pr(X_{n-1} = k \mid X_0 = i).$$

The first term of the last product equals $\Pr(X_n = j \mid X_{n-1} = k)$ (by the Markovian property), which is equal to $\mathbb{P}_{kj}$ (due to time-homogeneity). By the induction assumption, the second term equals $(\mathbb{P}^{n-1})_{ik}$. Putting these together, we get

$$\sum_k (\mathbb{P}^{n-1})_{ik}\,\mathbb{P}_{kj},$$

which corresponds to the matrix product of $\mathbb{P}^{n-1}$ and $\mathbb{P}$. The result thus equals $(\mathbb{P}^n)_{ij}$. $\qquad\qquad\square$

Example 2.6. (Refer to Example 2.1). If today is cloudy, what is the probability of its being rainy three days from now?

Solution. We must compute the (2nd, 3rd) elements of $\mathbb{P}^3$, or, more efficiently

$$
\begin{bmatrix} 0.4 & 0.5 & 0.1 \end{bmatrix}
\begin{bmatrix} 0.6 & 0.3 & 0.1 \\ 0.4 & 0.5 & 0.1 \\ 0.3 & 0.4 & 0.3 \end{bmatrix}
\begin{bmatrix} 0.1 \\ 0.1 \\ 0.3 \end{bmatrix}
= \begin{bmatrix} 0.4 & 0.5 & 0.1 \end{bmatrix}
\begin{bmatrix} 0.12 \\ 0.12 \\ 0.16 \end{bmatrix}
= 12.4\%.
$$

Note the initial/final state corresponds to the row/column of $\mathbb{P}^3$, respectively. This can be computed more easily by

$$
> \mathbb{P}_1 := \begin{bmatrix} 0.6 & 0.3 & 0.1 \\ 0.4 & 0.5 & 0.1 \\ 0.3 & 0.4 & 0.3 \end{bmatrix} :
$$

$$
> (\mathbb{P}_1)^3_{2,3};
$$

$$
0.1234
$$

$\square$

Similarly, if a *record* of several past states is given (such as Monday was sunny, Tuesday was sunny again, and Wednesday was cloudy), computing the probability of rainy on Saturday would yield the same answer (since we can *ignore all but the latest piece of information*).

Now we modify the question slightly: What is the probability of its being rainy on Saturday *and* Sunday? To answer this (labeling Monday as day 0), we first recall

$$
\Pr(A \cap B) = \Pr(A)\Pr(B \mid A) \Rightarrow \Pr(A \cap B \mid C) = \Pr(A \mid C)\Pr(B \mid A \cap C),
$$

which is the product rule, conditional upon C. Then we proceed as follows:

$$
\begin{aligned}
\Pr(X_5 = R \cap X_6 &= R \mid X_0 = S \cap X_1 = S \cap X_2 = C) \\
&= \Pr(X_5 = R \cap X_6 = R \mid X_2 = C) \\
&= \Pr(X_5 = R \mid X_2 = C)\Pr(X_6 = R \mid X_5 = R \cap X_2 = C) \\
&= \Pr(X_5 = R \mid X_2 = C)\Pr(X_6 = R \mid X_5 = R) = 0.124 \times 0.3 \\
&= 3.72\%.
\end{aligned}
$$

To summarize the basic rules of forecasting based on the past record:

1. Ignore all but the latest item of your record.
2. Given this, find the probability of reaching a specific state on the first day of your "forecast."
3. Given this state has been reached, take it to the next day of your forecast.
4. Continue until the last day of the forecast is reached.
5. Multiply all these probabilities.

If an *initial distribution* (say **d**, understood to be a one-column matrix) is given (for day 0), the probabilities of being in a given State n transitions later are given by the elements of

$$\mathbf{d}^{\mathrm{T}} \mathbb{P}^n,$$

where $\mathbf{d}^{\mathrm{T}}$ is the TRANSPOSE of **d** (making it a *one-row* matrix). The result is a one-row matrix of (final-state) probabilities.

Note when $\mathbb{P}$ is stochastic, $\mathbb{P}^n$ is too for any integer n (prove by induction – this rests on the fact a product of *any* two stochastic matrices, say $\mathbb{Q}$ and $\mathbb{P}$, is also stochastic, which can be proven by summing $\sum_k Q_{ik} P_{kj}$ over j).

2.3 LONG-RUN PROPERTIES

We now investigate the long-run development of FMCs, which is closely related to the behavior of $\mathbb{P}^n$ for large n. The simplest situation occurs when all elements of $\mathbb{P}$ are positive (a special case of the so-called REGULAR FMC, defined later).

One can show that in this case $\mathbb{P}^\infty = \lim_{n \to \infty} \mathbb{P}^n$ exists, and all of its rows are identical (this should be intuitively clear: the probability of a sunny day 10 years from now should be practically independent of the initial condition), that is,

$$\mathbb{P}^\infty = \begin{bmatrix} s_1 & s_2 & s_3 \\ s_1 & s_2 & s_3 \\ s_1 & s_2 & s_3 \end{bmatrix}$$

(for a three-state chain), where $\mathbf{s}^{\mathrm{T}} = [s_1, s_2, s_3]$ is called the STATIONARY DISTRIBUTION (the individual components are called stationary probabilities). Later, we will supply a formal proof of this, but let us look at the consequences first.

These probabilities have two interpretations; s_i represents

1. The probability of being in State i after many transitions (this limit is often reached in a handful of transitions);
2. The RELATIVE FREQUENCY of occurrence of State i in the long run (technically the limit of the relative frequency of occurrence when approaching an *infinite run*; again, in practice, a few hundred transitions is usually a good approximation).

By computing individual powers of the TPM for each of our four examples, one readily notices the first (weather) and the last (HT-type patterns) quickly converge to the type of matrix just described; in the latter case, this happens in one step:

$$> \mathbb{P}_1 := \begin{bmatrix} 0.6 & 0.3 & 0.1 \\ 0.4 & 0.5 & 0.1 \\ 0.3 & 0.4 & 0.3 \end{bmatrix} :$$

> **for** i **from** 3 **by** 3 **to** 9 **do**
  $evalm\left(\mathbb{P}_1^i\right);$
  end do:

$$\begin{bmatrix} 0.4900 & 0.3869 & 0.1240 \\ 0.4820 & 0.3940 & 0.1240 \\ 0.4700 & 0.3980 & 0.1320 \end{bmatrix}$$

$$\begin{bmatrix} 0.4844 & 0.3906 & 0.1250 \\ 0.4844 & 0.3906 & 0.1250 \\ 0.4842 & 0.3908 & 0.1251 \end{bmatrix}$$

$$\begin{bmatrix} 0.4844 & 0.3906 & 0.1250 \\ 0.4844 & 0.3906 & 0.1250 \\ 0.4844 & 0.3906 & 0.1250 \end{bmatrix}$$

$$> \mathbb{P}_4 := \begin{bmatrix} \frac{1}{2} & \frac{1}{2} & 0 & 0 \\ 0 & 0 & \frac{1}{2} & \frac{1}{2} \\ \frac{1}{2} & \frac{1}{2} & 0 & 0 \\ 0 & 0 & \frac{1}{2} & \frac{1}{2} \end{bmatrix} :$$

$> \mathbb{P}_4^2;$

$$\begin{bmatrix} \frac{1}{4} & \frac{1}{4} & \frac{1}{4} & \frac{1}{4} \\ \frac{1}{4} & \frac{1}{4} & \frac{1}{4} & \frac{1}{4} \\ \frac{1}{4} & \frac{1}{4} & \frac{1}{4} & \frac{1}{4} \\ \frac{1}{4} & \frac{1}{4} & \frac{1}{4} & \frac{1}{4} \end{bmatrix}$$

Knowing the special form of the limiting matrix, there is a shortcut to computing **s**: $\mathbb{P}^\infty \mathbb{P} = \mathbb{P}^\infty \Rightarrow \mathbf{s}^T \mathbb{P} = \mathbf{s}^T \Rightarrow \mathbb{P}^T \mathbf{s} = \mathbf{s} \Rightarrow (\mathbb{P}^T - \mathbb{I})\mathbf{s} = \mathbf{0}$. Solving the last set of (homogeneous) equations yields **s**. Since adding the elements of each row of $\mathbb{P} - \mathbb{I}$ results in **0**, the matrix (and its transpose) is singular, so there must be at least one nonzero solution to **s**. For regular FMCs, the solution is (up to a multiplicative constant) unique since the RANK of $\mathbb{P} - \mathbb{I}$ must equal $N - 1$, where N is the total number of possible states.

Example 2.7. Consider our weather example, where

$$\mathbb{P}^T - \mathbb{I} = \begin{bmatrix} -0.4 & 0.4 & 0.3 \\ 0.3 & -0.5 & 0.4 \\ 0.1 & 0.1 & -0.7 \end{bmatrix}.$$

The matrix is of rank 2; we may thus arbitrarily discard one of the three equations. Furthermore, since the solution can be determined up to a multiplicative constant, assuming s_3 is nonzero (as it must be in the regular case), we can set it to 1, eliminating one unknown. We then solve for s_1 and s_2 and multiply **s** by a constant, which makes it into a probability vector (we call this step NORMALIZING **s**). In terms of our example, we get

$$-0.4s_1 + 0.4s_2 = -0.3,$$
$$0.3s_1 - 0.5s_2 = -0.4.$$

The solution is given by

$$\begin{bmatrix} -0.5 & -0.4 \\ -0.3 & -0.4 \end{bmatrix} \begin{bmatrix} -0.3 \\ -0.4 \end{bmatrix} \frac{1}{0.08} = \begin{bmatrix} \frac{31}{8} \\ \frac{25}{8} \end{bmatrix},$$

together with $s_3 = 1$. Since, at this point, we do not care about the multiplicative factor, we may also present it as $[31, 25, 8]^T$ (the reader should verify this solution meets all *three* equations). Finally, since the final solution must correspond to a probability distribution (the components adding up to 1), all we need to do is normalize the answer, thus:

$$\mathbf{s} = \begin{bmatrix} \frac{31}{64} \\ \frac{25}{64} \\ \frac{8}{64} \end{bmatrix} = \begin{bmatrix} 0.4844 \\ 0.3906 \\ 0.1250 \end{bmatrix}.$$

And, true enough, this agrees with what we observed by taking large powers of the corresponding TPM   ◇

Example 2.8. Even though Example 2.3 is not regular (as we discover in the next section), it also has a unique solution to $\mathbf{s}^T \mathbb{P} = \mathbf{s}^T$. The solution is called the FIXED VECTOR, and it still corresponds to the relative frequencies of states in the long run (but no longer to the $\mathbb{P}^\infty$ limit). Finding this **s** is a bit more difficult now (we must solve a 5×5 set of equations), so let us see whether we can guess the answer. We conjecture that the proportion of time spent in each compartment is proportional to the number of doors to/from it. This would imply $\mathbf{s}^T$ should be proportional to $\begin{bmatrix} 1 & 2 & 1 & 2 & 3 & 1 \end{bmatrix}$, implying

$\mathbf{s}^T = \begin{bmatrix} \frac{1}{10} & \frac{2}{10} & \frac{1}{10} & \frac{2}{10} & \frac{3}{10} & \frac{1}{10} \end{bmatrix}$. To verify the correctness of this answer, we must check that

$$\begin{bmatrix} \frac{1}{10} & \frac{2}{10} & \frac{1}{10} & \frac{2}{10} & \frac{3}{10} & \frac{1}{10} \end{bmatrix} \begin{bmatrix} 0 & 0 & 0 & 1 & 0 & 0 \\ 0 & 0 & \frac{1}{2} & 0 & \frac{1}{2} & 0 \\ 0 & 1 & 0 & 0 & 0 & 0 \\ \frac{1}{2} & 0 & 0 & 0 & \frac{1}{2} & 0 \\ 0 & \frac{1}{3} & 0 & \frac{1}{3} & 0 & \frac{1}{3} \\ 0 & 0 & 0 & 0 & 1 & 0 \end{bmatrix}$$

equals $\begin{bmatrix} \frac{1}{10} & \frac{2}{10} & \frac{1}{10} & \frac{2}{10} & \frac{3}{10} & \frac{1}{10} \end{bmatrix}$, which is indeed the case. But this time

$$\mathbb{P}^{100} \approx \mathbb{P}^{102} \approx \cdots \approx \begin{bmatrix} 0.2 & 0 & 0.2 & 0 & 0.6 & 0 \\ 0 & 0.4 & 0 & 0.4 & 0 & 0.2 \\ 0.2 & 0 & 0.2 & 0 & 0.6 & 0 \\ 0 & 0.4 & 0 & 0.4 & 0 & 0.2 \\ 0.2 & 0 & 0.2 & 0 & 0.6 & 0 \\ 0 & 0.4 & 0 & 0.4 & 0 & 0.2 \end{bmatrix}$$

and

$$\mathbb{P}^{101} \approx \mathbb{P}^{103} \approx \cdots \approx \begin{bmatrix} 0 & 0.4 & 0 & 0.4 & 0 & 0.2 \\ 0.2 & 0 & 0.2 & 0 & 0.6 & 0 \\ 0 & 0.4 & 0 & 0.4 & 0 & 0.2 \\ 0.2 & 0 & 0.2 & 0 & 0.6 & 0 \\ 0 & 0.4 & 0 & 0.4 & 0 & 0.2 \\ 0.2 & 0 & 0.2 & 0 & 0.6 & 0 \end{bmatrix}$$

(there appear to be two alternating limits). In the next section, we explain why. $\diamond$

Example 2.9. Recalling Example 2.4 we can easily gather that each of the four patterns (HH, HT, TH, and HH) must have the same frequency of occurrence, and the stationary probabilities should thus all equal $\frac{1}{4}$ each. This can be verified by

$$\begin{bmatrix} \frac{1}{4} & \frac{1}{4} & \frac{1}{4} & \frac{1}{4} \end{bmatrix} \begin{bmatrix} \frac{1}{2} & \frac{1}{2} & 0 & 0 \\ 0 & 0 & \frac{1}{2} & \frac{1}{2} \\ \frac{1}{2} & \frac{1}{2} & 0 & 0 \\ 0 & 0 & \frac{1}{2} & \frac{1}{2} \end{bmatrix} = \begin{bmatrix} \frac{1}{4} & \frac{1}{4} & \frac{1}{4} & \frac{1}{4} \end{bmatrix}.$$

And, sure enough,

$$\begin{bmatrix} \frac{1}{2} & \frac{1}{2} & 0 & 0 \\ 0 & 0 & \frac{1}{2} & \frac{1}{2} \\ \frac{1}{2} & \frac{1}{2} & 0 & 0 \\ 0 & 0 & \frac{1}{2} & \frac{1}{2} \end{bmatrix}^n = \begin{bmatrix} \frac{1}{4} & \frac{1}{4} & \frac{1}{4} & \frac{1}{4} \\ \frac{1}{4} & \frac{1}{4} & \frac{1}{4} & \frac{1}{4} \\ \frac{1}{4} & \frac{1}{4} & \frac{1}{4} & \frac{1}{4} \\ \frac{1}{4} & \frac{1}{4} & \frac{1}{4} & \frac{1}{4} \end{bmatrix} \quad \text{for } n = 2, 3, \dots,$$

as we already discovered through MAPLE. $\diamond$

Example 2.10. Computing individual powers of $\mathbb{P}$ from Example 2.2, we can establish the limit (reached, to a good accuracy, only at $\mathbb{P}^{30}$) is

$$\begin{bmatrix} 1 & 0 & 0 & 0 & 0 \\ \frac{3}{4} & 0 & 0 & 0 & \frac{1}{4} \\ \frac{1}{2} & 0 & 0 & 0 & \frac{1}{2} \\ \frac{1}{4} & 0 & 0 & 0 & \frac{3}{4} \\ 0 & 0 & 0 & 0 & 1 \end{bmatrix}.$$

Now, even though the $\mathbb{P}^\infty$ limit exists, it has a totally different structure than in the regular case. $\diamond$

So there are several questions we would like to resolve:

1. How do we know (without computing $\mathbb{P}^\infty$) that an FMC is regular?
2. When does a TPM have a fixed vector but not a stationary distribution, and what is the pattern of large powers of $\mathbb{P}$ in such a case?
3. What else can happen to $\mathbb{P}^\infty$ (in the nonregular cases), and how do we find this without computing high powers of $\mathbb{P}$?

To sort out all these questions and discover the full story of the long-run behavior of an FMC, a brand new approach is called for.

2.4 CLASSIFICATION OF STATES

A DIRECTED GRAPH of a TPM is a diagram in which each state is represented by a small circle, and each potential (nonzero) transition by a directed arrow.

Example 2.11. A directed graph based on the TPM of Example 2.1.

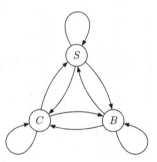

◇

Example 2.12. A directed graph based on the TPM of Example 2.2.   ◇

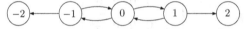

Example 2.13. A directed graph based on the TPM of Example 2.3.   ◇

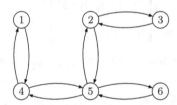

Example 2.14. A directed graph based on the TPM of Example 2.4.

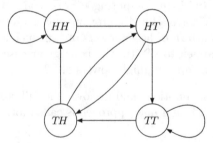

◇

From such a graph one can gather much useful information about the corresponding FMC.

A natural question to ask about any two states, say a and b, is this: Is it possible to get from a to b in some (including 0) number of steps, and then, similarly, from b to a? If the answer is YES (to *both*), we say a and b COMMUNICATE (and denote this by $a \leftrightarrow b$).

Mathematically, a RELATION assigns each (ordered) pair of elements (states, in our case) a YES or NO value. A relation (denoted by $a \to b$ in general) can be symmetric ($a \to b \Rightarrow b \to a$), antisymmetric ($a \to b \Rightarrow \neg(b \to a)$), reflexive ($a \to a$ for each a), or transitive ($a \to b$ and $b \to c \Rightarrow a \to c$).

Is our "communicate" relation symmetric? (YES). Antisymmetric? (NO). Reflexive? (YES, that is why we said "including 0"). Transitive? (YES). A relation that is symmetric, reflexive, and transitive is called an EQUIVALENCE RELATION (a relation that is antisymmetric, reflexive, and transitive is called a PARTIAL ORDER).

An equivalence relation implies we can subdivide the original set (of states, in our case) into so-called EQUIVALENCE CLASSES (each state will be a member of exactly one such class; the classes are thus mutually exclusive, and their union covers the whole set – no gaps, no overlaps). To find these classes, we start with an arbitrary state (say a) and collect all states that communicate with a (together with a, these constitute Class 1); then we take, arbitrarily, any state outside Class 1 (say State b) and find all states that communicate with b (this will be Class 2), and so on till the (finite) set is exhausted.

Example 2.15. Our first, third, and fourth examples each consist of a single class of states (all states communicate with one another). In the second example, States -1, 0, and 1 communicate with one another (one class), but there is no way to reach any other state from State 2 (a class by itself) and also form State -2 (the last class). $\diamondsuit$

In a more complicated situation, it helps to look for closed loops (all states along a closed loop communicate with one another; if, for example, two closed loops have a common element, then they must both belong to the same class).

Once we partition our states into classes, what is the relationship among the classes themselves? It may still be possible to move from one class to another (but not back), so some classes will be connected by one-directional arrows (defining a relationship between *classes* – this relationship is, by definition, antisymmetric, reflexive, and transitive; the reflexive property means the class is connected to itself). Note this time there can be no closed loops – they would create a single class. Also note two classes being connected (say $A \to B$) implies we can get from *any* state of Class A to *any* state of Class B.

It is also possible some classes (or set of classes) are totally disconnected from the rest (no connection in either direction). In practice, this can happen only when we combine two FMCs, which have nothing to do with one another, into a single FMC – using matrix notation, something like this:

$$\begin{bmatrix} \mathbb{P}_1 & \mathbb{O} \\ \mathbb{O} & \mathbb{P}_2 \end{bmatrix},$$

where $\mathbb{O}$ represents a zero matrix. So should this happen to us, we can investigate each disconnected group of classes on its own, ignoring the rest (i.e., why this hardly ever happens – it would be mathematically trivial and practically meaningless).

There are two important definitions relating to classes (and their one-way connections): a class that *cannot be left* (found at the bottom of a connection diagram, if all arrows point down) is called RECURRENT; any other class (with an *outgoing arrow*) is called TRANSIENT (these terms are also applied to individual *states* inside these classes). We will soon discover that ultimately (in the long run), an FMC must end up in one of the recurrent classes (the probability of staying transient indefinitely is zero). Note we cannot have transient classes alone (there must always be at least one recurrent class). On the other hand, an FMC can consist of recurrent classes only (normally, only one; see the discussion of the previous paragraph).

We mention in passing that all EIGENVALUES of a TPM must be, in absolute value, less than or equal to 1. One of these eigenvalues *must* be equal to 1, and its MULTIPLICITY yields the number of *recurrent* classes of the corresponding FMC.

Example 2.16. Consider the TPM from Example 2.2.

$$> \mathbb{P}_2 := \begin{bmatrix} 1 & 0 & 0 & 0 & 0 \\ \frac{1}{2} & 0 & \frac{1}{2} & 0 & 0 \\ 0 & \frac{1}{2} & 0 & \frac{1}{2} & 0 \\ 0 & 0 & \frac{1}{2} & 0 & \frac{1}{2} \\ 0 & 0 & 0 & 0 & 1 \end{bmatrix} :$$

$> evalf\,(Eigenvalues\,(\mathbb{P}_2, output = list))\,;$

$$[0.0,\ 0.7071,\ -0.7071,\ 1.0000,\ 1.0000]$$

indicating the presence of *two* recurrent classes.   ◇

After we have partitioned an FMC into classes, it is convenient to relabel the individual states (and correspondingly rearrange the TPM), so states of the same class are consecutive (the TPM is then organized into so-called BLOCKS), *starting* with recurrent classes; try to visualize what the complete TPM will then look like. Finally, $\mathbb{P}$ can be divided into four basic superblocks by separating the recurrent and transient parts only (never mind the individual classes); thus:

$$\mathbb{P} = \begin{bmatrix} \mathbb{R} & \mathbb{O} \\ \mathbb{U} & \mathbb{T} \end{bmatrix},$$

where $\mathbb{O}$ again denotes the zero matrix (there are no transitions from recurrent to transient states). It is easy to show

$$\mathbb{P}^n = \begin{bmatrix} \mathbb{R}^n & \mathbb{O} \\ ? & \mathbb{T}^n \end{bmatrix}$$

(with the lower left superblock being somehow more complicated). This already greatly simplifies our task of figuring out what happens to large powers of $\mathbb{P}$.

Proposition 2.2.

$$\mathbb{T}^\infty \to \mathbb{O},$$

meaning transient states, in the long run, disappear – the FMC must eventually enter one of its recurrent classes and stay there for good since there is no way out.

Proof. Let $P_a^{(k)}$ be the probability that, starting from a transient state a, k transitions later we will have already reached a recurrent class. These probabilities are nondecreasing in k (once recurrent, always recurrent). The fact it is *possible* to reach a recurrent class from any a (transient) effectively means this: for each a there is a number of transitions, say k_a, such that $P_a^{(k)}$ is already positive, say p_a. If we now take the largest of these k_a (say K) and the smallest of the p_a (say p), then we conclude $P_a^{(K)} \geq p$ for each a (transient), or, equivalently, $Q_a^{(K)} < 1 - p$ (where Q_a^k is the probability that a has not yet left left the transient states after K transitions, that is, $\sum_{b \in T}(\mathbb{P}^k)_{ab}$, where T is the set of all transient states).

Now,

$$Q_a^{(2K)} = \sum_{b \in T}(\mathbb{P}^{2k})_{ab}$$

$$= \sum_{b \in T}\sum_c (\mathbb{P}^k)_{ac}(\mathbb{P}^k)_{cb}$$

(the c summation is over *all* states)

$$\leq \sum_{b \in T}\sum_{c \in T}(\mathbb{P}^k)_{ac}(\mathbb{P}^k)_{cb}$$

$$\leq (1 - p)\sum_{c \in T}(\mathbb{P}^k)_{ac}$$

$$\leq (1 - p)^2.$$

Similarly, one can show

$$Q_a^{(3K)} \leq (1-p)^3,$$
$$Q_a^{(4K)} \leq (1-p)^4,$$
$$\vdots \qquad \vdots$$

implying $Q_a^{(\infty)} \leq \lim_{n \to \infty} (1-p)^n = 0$. This shows the probability that a transient state a stays transient indefinitely is zero. Thus, every transient state is eventually captured in one of the recurrent classes, with probability of 1.   □

Next we tackle the upper left corner of $\mathbb{P}^n$. First of all, $\mathbb{R}$ itself breaks down into individual classes, thus:

$$\begin{bmatrix} \mathbb{R}_1 & \mathbb{O} & \cdots & \mathbb{O} \\ \mathbb{O} & \mathbb{R}_2 & \cdots & \mathbb{O} \\ \vdots & \vdots & \ddots & \vdots \\ \mathbb{O} & \mathbb{O} & \cdots & \mathbb{R}_k \end{bmatrix}$$

since recurrent classes do not communicate (not even one way). Clearly, then,

$$\mathbb{R}^n = \begin{bmatrix} \mathbb{R}_1^n & \mathbb{O} & \cdots & \mathbb{O} \\ \mathbb{O} & \mathbb{R}_2^n & \cdots & \mathbb{O} \\ \vdots & \vdots & \ddots & \vdots \\ \mathbb{O} & \mathbb{O} & \cdots & \mathbb{R}_k^n \end{bmatrix},$$

and to find out what happens to this matrix for large n, we need to understand what happens to each of the $\mathbb{R}_i^n$ individually. We can thus restrict our attention to a *single* recurrent class.

To be able to fully understand the behavior of any such $\mathbb{R}_i^n$ (for large n), we first need to have a closer look *inside* the recurrent class, discovering a finer structure: a division into periodic *subclasses*.

2.5 Periodicity of a Class

Let us consider a single recurrent class (an FMC in its own right). If k_1, k_2, k_3, ... is a *complete* (and therefore infinite) list of the *number of transitions* in which one can return to the initial state (say a) – note this information can be gathered from the corresponding directed graph – then the *greatest common divisor* (say λ) of this set of integers is called the PERIOD of State a.

This can be restated as follows. If the length of every possible closed loop passing (at least once) through State a is divisible by λ, and if λ is the greatest of all integers for which this is true, then λ is the corresponding period. Note a closed loop is allowed any amount of duplication (both in terms of states and transitions) – we can go through the same loop, repeatedly, as many times as we like.

The last definition gives the impression that each state may have its own period. This is not the case.

Proposition 2.3. *The value of λ is the same regardless of which state is chosen for a. The period is thus a property of the whole class.*

Proof. Suppose State a has a period λ_a and State b has a (potentially different) period, say λ_b. Every closed loop passing through b either already passes through a or else can be easily extended (by a $b \to a \to b$ loop) to do so. Either way, the length of the loop must be divisible by λ_a (the extended loop is divisible by λ_a, and the extension itself is also divisible by λ_a; therefore, the difference between the two must be divisible by λ_a). This proves $\lambda_b \geq \lambda_a$. We can now reverse the argument and prove $\lambda_a \geq \lambda_b$, implying $\lambda_a = \lambda_b$. □

In practice, we just need to find the greatest common divisor of all closed loops found in the corresponding directed graph. Whenever there is a loop of length one (a state returning back to itself), the period must be equal to 1 (the class is then called *aperiodic* or *regular*). The same is true whenever we find one closed loop of length 2 and another of length 3 (or any other prime numbers). One should also keep in mind the period cannot be higher than the total number of states (thus, the number of possibilities is quite limited).

A trivial example of a class with a period equal to λ would be a simple CYCLE of λ states, where State 1 goes (in one transition) only to State 2, which in turn must go to State 3, etc., until State λ transits back to State 1 (visualize the directed graph). However, most periodic classes are more complicated than this!

The implication of a nontrivial (> 1) period λ is that we can further partition the set of states into λ SUBCLASSES, which are found as follows.

1. Select an arbitrary State a. It will be a member of Subclass 0 (we will label the subclasses 0, 1, 2, ..., $\lambda - 1$).
2. Find a path that starts at a and visits all states (some more than once if necessary).
3. Assign each state along this path to Subclass k mod λ, where k is the number of transitions to reach it.

It is quite simple to realize this definition of subclasses is consistent (each state is assigned to the same subclass no matter how many times we go through it) and, up to a cyclical rearrangement, unique (we get the same answer regardless of where we start and which path we choose).

Note subclasses do not need to be of the same size!

Example 2.17. Find the subclasses of the following FMC (defined by the corresponding TMP):

$$
\begin{bmatrix}
0 & 0.7 & 0 & 0 & 0.3 \\
0.7 & 0 & 0 & 0.3 & 0 \\
0.5 & 0 & 0 & 0.5 & 0 \\
0 & 0.2 & 0.2 & 0 & 0.6 \\
0.4 & 0 & 0 & 0.6 & 0
\end{bmatrix} .
$$

Solution. From the corresponding directed graph

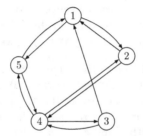

it follows this *is* a *single* class (automatically recurrent). Since State 1 can go to State 5 and then back to State 1, there is a closed loop of length 2 (the period cannot be any higher, that is, it must be either 2 or 1). Since all closed loops we find in the directed graph are of length 2 or 4 (and higher multiples of 2), the period is equal to 2. From the path $1 \to 5 \to 4 \to 3 \to 4 \to 2$ we can conclude the two subclasses are $\{1, 4\}$ and $\{2, 3, 5\}$. Rearranging our TPM accordingly we get

	1	4	2	3	5
1	0	0	0.7	0	0.3
4	0	0	0.2	0.2	0.6
2	0.7	0.3	0	0	0
3	0.5	0.5	0	0	0
5	0.4	0.6	0	0	0

$$(2.1)$$

Note this partitions the matrix into corresponding subblocks.   ◇

One can show the last observation is true in general, that is, one can go (in one transition) only from Subclass 0 to Subclass 1, from Subclass 1 to Subclass 2, etc., until finally, one goes from Subclass $\lambda - 1$ back to Subclass 0. The rearranged TPM will then always look like this (we use a hypothetical example with four subclasses):

$$
\mathbb{R} = \begin{bmatrix}
\mathbb{O} & \mathbb{C}_1 & \mathbb{O} & \mathbb{O} \\
\mathbb{O} & \mathbb{O} & \mathbb{C}_2 & \mathbb{O} \\
\mathbb{O} & \mathbb{O} & \mathbb{O} & \mathbb{C}_3 \\
\mathbb{C}_4 & \mathbb{O} & \mathbb{O} & \mathbb{O}
\end{bmatrix},
$$

where the size of each subblock corresponds to the number of states in the respective (row and column) subclasses.

Note $\mathbb{R}^\lambda$ will be (block) diagonal; for our last example, this means

$$
\mathbb{R}^4 = \begin{bmatrix}
\mathbb{C}_1\mathbb{C}_2\mathbb{C}_3\mathbb{C}_4 & \mathbb{O} & \mathbb{O} & \mathbb{O} \\
\mathbb{O} & \mathbb{C}_2\mathbb{C}_3\mathbb{C}_4\mathbb{C}_1 & \mathbb{O} & \mathbb{O} \\
\mathbb{O} & \mathbb{O} & \mathbb{C}_3\mathbb{C}_4\mathbb{C}_1\mathbb{C}_2 & \mathbb{O} \\
\mathbb{O} & \mathbb{O} & \mathbb{O} & \mathbb{C}_4\mathbb{C}_1\mathbb{C}_2\mathbb{C}_3
\end{bmatrix}
$$

(from this one should be able to discern the general pattern). Note by taking four transitions at a time (seen as a single supertransition), the process turns into an FMC with four recurrent *classes* (no longer subclasses), which we know how to deal with. This implies $\lim_{n \to \infty} \mathbb{R}^{4n}$ will have the following form:

$$
\begin{bmatrix}
\mathbb{S}_1 & \mathbb{O} & \mathbb{O} & \mathbb{O} \\
\mathbb{O} & \mathbb{S}_2 & \mathbb{O} & \mathbb{O} \\
\mathbb{O} & \mathbb{O} & \mathbb{S}_3 & \mathbb{O} \\
\mathbb{O} & \mathbb{O} & \mathbb{O} & \mathbb{S}_4
\end{bmatrix},
$$

where $\mathbb{S}_1$ is a matrix with identical rows, say s_1 (the stationary probability vector of $\mathbb{C}_1\mathbb{C}_2\mathbb{C}_3\mathbb{C}_4$ - one can show each of the four new classes must be *aperiodic*); similarly, $\mathbb{S}_2$ consists of stationary probabilities s_2 of $\mathbb{C}_2\mathbb{C}_3\mathbb{C}_4\mathbb{C}_1$ etc.

What happens when the process undergoes one extra transition? This is quite simple: it goes from Subclass 0 to Subclass 1, (or $1 \to 2$ or $2 \to 3$ or $3 \to 0$), but the probabilities within each subclass must remain stationary. This is clear from the following limit:

$$\lim_{n\to\infty} \mathbb{R}^{4n+1} = \left(\lim_{n\to\infty} \mathbb{R}^{4n}\right)\mathbb{R}$$

$$= \begin{bmatrix} \mathbb{O} & S_1C_1 & \mathbb{O} & \mathbb{O} \\ \mathbb{O} & \mathbb{O} & S_2C_2 & \mathbb{O} \\ \mathbb{O} & \mathbb{O} & \mathbb{O} & S_3C_3 \\ S_4C_4 & \mathbb{O} & \mathbb{O} & \mathbb{O} \end{bmatrix}$$

$$= \begin{bmatrix} \mathbb{O} & S_2 & \mathbb{O} & \mathbb{O} \\ \mathbb{O} & \mathbb{O} & S_3 & \mathbb{O} \\ \mathbb{O} & \mathbb{O} & \mathbb{O} & S_4 \\ S_1 & \mathbb{O} & \mathbb{O} & \mathbb{O} \end{bmatrix}$$

since $s_1^T C_1$ is a solution to $s^T C_2 C_3 C_4 C_1 = s^T$ (note s_1 satisfies $s_1^T C_1$ $C_2 C_3 C_4 = s_1^T$) and must therefore be equal to s_2. Similarly, $s_2^T C_2 = s_3^T$, $s_3^T C_3 = s_4^T$, and $s_4^T C_4 = s_1^T$ (back to s_1).

This implies once we obtain *one* of the s vectors, we can get the rest by a simple multiplication. We would like to start from the *shortest* one since it is the easiest to find.

The fixed vector of $\mathbb{R}$ (a solution to $f^T \mathbb{R} = f^T$) is then found by

$$f^T = \frac{(s_1^T, s_2^T, s_3^T, s_4^T)}{4},$$

and similarly for any other number of subclasses. The interpretation is clear: this yields the long-run proportion of visits to individual states of the class.

Example 2.18. Returning to our two subclasses of Example 2.17, we first compute

$$C_1 C_2 = \begin{bmatrix} 0.7 & 0 & 0.3 \\ 0.2 & 0.2 & 0.6 \end{bmatrix}\begin{bmatrix} 0.7 & 0.3 \\ 0.5 & 0.5 \\ 0.4 & 0.6 \end{bmatrix} = \begin{bmatrix} 0.61 & 0.39 \\ 0.48 & 0.52 \end{bmatrix},$$

then find the corresponding $s_1 = \begin{bmatrix} \frac{16}{29} \\ \frac{13}{29} \end{bmatrix}$ (a relatively simple exercise), and finally

$$s_2^T = \begin{bmatrix} \frac{16}{29} & \frac{13}{29} \end{bmatrix}\begin{bmatrix} \frac{7}{10} & 0 & \frac{3}{10} \\ \frac{2}{10} & \frac{2}{10} & \frac{6}{10} \end{bmatrix} = \begin{bmatrix} \frac{69}{145} & \frac{13}{145} & \frac{63}{145} \end{bmatrix}.$$

To verify the two answers, we can now build the (unique) fixed probability vector of the original PTM, namely, $\mathbf{f}^T = \begin{bmatrix} \frac{16}{58} & \frac{13}{58} & \frac{69}{290} & \frac{13}{290} & \frac{63}{290} \end{bmatrix}$, and check $\mathbf{f}^T = \mathbb{R}\mathbf{f}^T$ (which is indeed the case). $\diamond$

Similarly to the previous $\lim\limits_{n\to\infty} \mathbb{R}^{4n+1}$, we can derive

$$\lim_{n\to\infty} \mathbb{R}^{4n+2} = \begin{bmatrix} \mathbb{O} & \mathbb{O} & \mathbb{S}_3 & \mathbb{O} \\ \mathbb{O} & \mathbb{O} & \mathbb{O} & \mathbb{S}_4 \\ \mathbb{S}_1 & \mathbb{O} & \mathbb{O} & \mathbb{O} \\ \mathbb{O} & \mathbb{S}_2 & \mathbb{O} & \mathbb{O} \end{bmatrix} \quad \text{and} \quad \lim_{n\to\infty} \mathbb{R}^{4n+3} = \begin{bmatrix} \mathbb{O} & \mathbb{O} & \mathbb{O} & \mathbb{S}_4 \\ \mathbb{S}_1 & \mathbb{O} & \mathbb{O} & \mathbb{O} \\ \mathbb{O} & \mathbb{S}_2 & \mathbb{O} & \mathbb{O} \\ \mathbb{O} & \mathbb{O} & \mathbb{S}_3 & \mathbb{O} \end{bmatrix},$$

where $\mathbb{S}_i$ implies a matrix whose rows are all equal to $\mathbf{s}_i$ (but its row dimension may change from one power of $\mathbb{R}$ to the next).

So now we know how to raise $\mathbb{R}$ to any *large* power.

Example 2.19. Find

$$\begin{bmatrix} \cdot & \cdot & \cdot & 0.2 & 0.8 & \cdot & \cdot \\ \cdot & \cdot & \cdot & 0.5 & 0.5 & \cdot & \cdot \\ \cdot & \cdot & \cdot & 1 & 0 & \cdot & \cdot \\ \cdot & \cdot & \cdot & \cdot & \cdot & 0.7 & 0.3 \\ \cdot & \cdot & \cdot & \cdot & \cdot & 0.4 & 0.6 \\ 0.3 & 0.2 & 0.5 & \cdot & \cdot & \cdot & \cdot \\ 0.2 & 0 & 0.8 & \cdot & \cdot & \cdot & \cdot \end{bmatrix}^{10000}$$

(dots represent zeros).

Solution. One can confirm the period of the corresponding class is 3, and the subclasses are $\{1,2,3\}$, $\{4,5\}$ and $\{6,7\}$. To get the stationary probabilities of the second subclass, we first need

$$\mathbb{C}_2\mathbb{C}_3\mathbb{C}_1 = \begin{bmatrix} 0.7 & 0.3 \\ 0.4 & 0.6 \end{bmatrix} \begin{bmatrix} 0.3 & 0.2 & 0.5 \\ 0.2 & 0 & 0.8 \end{bmatrix} \begin{bmatrix} 0.2 & 0.8 \\ 0.5 & 0.5 \\ 1 & 0 \end{bmatrix} = \begin{bmatrix} 0.714 & 0.286 \\ 0.768 & 0.232 \end{bmatrix}$$

whose stationary vector is $s_2^T = \begin{bmatrix} 0.72865 & 0.27135 \end{bmatrix}$ (verify). Then

$$s_3^T = s_2^T \mathbb{C}_2 = \begin{bmatrix} 0.61860 & 0.38140 \end{bmatrix}$$

and

$$s_1^T = s_3^T \mathbb{C}_3 = \begin{bmatrix} 0.26186 & 0.12372 & 0.61442 \end{bmatrix}.$$

Since $10000 \bmod 3 \equiv 1$, the answer is

$$\begin{bmatrix} \cdot & \cdot & \cdot & 0.72865 & 0.27135 & \cdot & \cdot \\ \cdot & \cdot & \cdot & 0.72865 & 0.27135 & \cdot & \cdot \\ \cdot & \cdot & \cdot & 0.72865 & 0.27135 & \cdot & \cdot \\ \cdot & \cdot & \cdot & \cdot & \cdot & 0.61860 & 0.38140 \\ \cdot & \cdot & \cdot & \cdot & \cdot & 0.61860 & 0.38140 \\ 0.26186 & 0.12372 & 0.61442 & \cdot & \cdot & \cdot & \cdot \\ 0.26186 & 0.12372 & 0.61442 & \cdot & \cdot & \cdot & \cdot \end{bmatrix}$$

Similarly, the 10001th power of the original matrix would be equal to

$$\begin{bmatrix} \cdot & \cdot & \cdot & \cdot & \cdot & 0.61860 & 0.38140 \\ \cdot & \cdot & \cdot & \cdot & \cdot & 0.61860 & 0.38140 \\ \cdot & \cdot & \cdot & \cdot & \cdot & 0.61860 & 0.38140 \\ 0.26186 & 0.12372 & 0.61442 & \cdot & \cdot & \cdot & \cdot \\ 0.26186 & 0.12372 & 0.61442 & \cdot & \cdot & \cdot & \cdot \\ \cdot & \cdot & \cdot & 0.72865 & 0.27135 & \cdot & \cdot \\ \cdot & \cdot & \cdot & 0.72865 & 0.27135 & \cdot & \cdot \end{bmatrix}$$

At this point it should be clear what the 10002th power looks like.   ◇

Remark 2.1. A recurrent class with a period of λ contributes all λ roots of 1 (each exactly once) to the eigenvalues of the corresponding TPM (the remaining eigenvalues must be, in absolute value, less than 1). Thus, eigenvalues nicely reveal the number and periodicity of all recurrent classes.

$$> \mathbb{T} := \begin{bmatrix} 0 & 0 & 0 & .2 & .8 & 0 & 0 \\ 0 & 0 & 0 & 0.5 & 0.5 & 0 & 0 \\ 0 & 0 & 0 & 1 & 0 & 0 & 0 \\ 0 & 0 & 0 & 0 & 0 & 0.7 & 0.3 \\ 0 & 0 & 0 & 0 & 0 & 0.4 & 0.6 \\ 0.3 & 0.2 & 0.5 & 0 & 0 & 0 & 0 \\ 0.2 & 0 & 0.8 & 0 & 0 & 0 & 0 \end{bmatrix} :$$

$>$ $\mathbb{T} := convert\,(\mathbb{T}, rational)$: {we do this to get exact eigenvalues}
$>$ $\lambda := Eigenvalues(\mathbb{T}, output = list);$

$$\lambda := \Big[\; 0, \quad 1, \quad -1/2 - 1/2\,\mathbf{I}\sqrt{3}, \quad -1/2 + 1/2\,\mathbf{I}\sqrt{3},$$

$$-3/10\,\sqrt[3]{2}, \quad \tfrac{3}{20}\,\sqrt[3]{2} - \tfrac{3}{20}\,\mathbf{I}\sqrt{3}\,\sqrt[3]{2}, \quad \tfrac{3}{20}\,\sqrt[3]{2} + \tfrac{3}{20}\,\mathbf{I}\sqrt{3}\,\sqrt[3]{2} \;\Big]$$

$>$ $seq\,(evalf\,(abs\,(x))\,, x \in \lambda)\,;$

$$0.000, 1.0000, 1.0000, 1.0000, 0.3780, 0.3780, 0.3780$$

$>$ $simplify\,\big(seq\,(\lambda_i^3, i \in [2, 3, 4])\big)\,;$

$$1, 1, 1$$

This implies there is only a single recurrent class whose period is 3.

2.6 REGULAR MARKOV CHAINS

An FMC with a single, aperiodic class is called REGULAR. We already know that for these, $\mathbb{P}^\infty$ exists and has a stationary vector in each row. We can prove this in four steps (three propositions and a conclusion).

Proposition 2.4. *If S is a nonempty set of positive integers closed under addition and having 1 as its greatest common divisor, then starting from a certain integer, say N, all integers $(\geq N)$ must be in S.*

Proof. We know (from number theory) there must be a finite set of integers from S (we call them $n_1, n_2, \ldots, n_k$) whose linear combination (with integer coefficients $a_1, a_2, \ldots, a_k$) must be equal to the corresponding greatest common divisor; thus,

$$a_1 n_1 + a_2 n_2 + \cdots + a_k n_k = 1.$$

Collecting the positive and negative terms on the left-hand side of this equation implies

$$N_1 - N_2 = 1,$$

where both N_1 and N_2 belong to S (due to its closure under addition). Let q be any integer $\geq N_2(N_2 - 1)$. Since q can be written as $a N_2 + b$, where $0 \leq b < N_2$ and $a \geq N_2 - 1$, and since

$$a N_2 + b = (a - b)N_2 + b(1 + N_2) = (a - b)N_2 + bN_1,$$

each such q must be a member of S (again, due to the closure property).   □

Proposition 2.5. *The set of integers n for which $(\mathbb{P}^n)_{ii} > 0$, where $\mathbb{P}$ is regular, is closed under addition for each i. This implies, for sufficiently large n, all elements of $\mathbb{P}^N$ are strictly positive (meaning it is possible to move from State i back to State i in exactly n transitions).*

Proof. Since

$$(\mathbb{P}^{n+m})_{ij} = \sum_k (\mathbb{P}^n)_{ik}(\mathbb{P}^m)_{kj} \geq (\mathbb{P}^n)_{ii}(\mathbb{P}^m)_{ij} > 0,$$

where m is smaller than the total number of states (since State j can be reached from State i by visiting any of the other states no more than once). We can thus see, for sufficiently large n, all $\mathbb{P}^n_{ij}$ are strictly positive (i.e., have no zero entries).   □

When a stochastic matrix $\mathbb{P}$ multiplies a column vector $\mathbf{r}$, each component of the result is a (different) weighted average of the elements of $\mathbf{r}$. The smallest value of $\mathbb{P}\mathbf{r}$ thus cannot be any smaller than that of $\mathbf{r}$ (similarly, the largest value cannot go up). We now take $\mathbb{Q} = \mathbb{P}^N$, where $\mathbb{P}$ is regular and N is large enough to eliminate zero entries from $\mathbb{Q}$. Clearly, there must be a positive ε such that all entries of $\mathbb{Q}$ are $\geq \varepsilon$. This implies the difference between the largest and smallest component of $\mathbb{Q}\mathbf{r}$ (let us call them M_1 and m_1, respectively) must be *smaller* than the difference between the largest and smallest components of $\mathbf{r}$ (let us call these M_0 and m_0) by a factor of at least $(1 - 2\varepsilon)$.

Proposition 2.6.

$$\max(\mathbb{Q}\mathbf{r}) - \min(\mathbb{Q}\mathbf{r}) \leq (1 - 2\varepsilon)\max(\mathbf{r}) - \min(\mathbf{r}).$$

Proof. Clearly, $m_1 \geq \varepsilon M_0 + (1-\varepsilon)m_0$ if we try to make the right-hand side as small as possible (multiplying M_0 by the smallest possible value and making all the other entries of $\mathbf{r}$ as small as possible). Similarly, $M_1 \leq \varepsilon m_0 + (1-\varepsilon)M_0$ (now we are multiplying m_0 by the smallest possible factor, leaving the rest for M_0). Subtracting the two inequalities yields

$$M_1 - m_1 \leq (1 - 2\varepsilon)(M_0 - m_0).$$

□

Proposition 2.7. *All rows of* $\mathbb{P}^{\infty}$ *are identical and equal to the stationary vector.*

Proof. Take $\mathbf{r}_1$ to be a *column* vector defined by $[1, 0, 0, \ldots, 0]^{\mathrm{T}}$ and multiply it repeatedly, say n times, by $\mathbb{Q}$, getting $\mathbb{Q}^n \mathbf{r}_1$ (the first column of $\mathbb{Q}^n$). The difference between the largest and smallest elements of the resulting vector is no bigger than $(1 - 2\varepsilon)^n$ – the previous proposition, applied n times – and converges to 0 when $n \to \infty$. Similarly (using the original $\mathbb{P}$) the difference between the largest and smallest elements of $\mathbb{P}^n \mathbf{r}_1$ must converge to 0 since it is a nonincreasing sequence that contains a subsequence (that, coming from $\mathbb{Q}^n \mathbf{r}_1$) converging to 0. We have thus proved the first column of $\mathbb{P}^n$ converges to a vector with constant elements. By taking $\mathbf{r}_2 = [0, 1, 0, \ldots, 0]^{\mathrm{T}}$ we can prove the same thing for each column of $\mathbb{P}^n$.   □

2.A INVERTING MATRICES

INVERTING (SMALL) MATRICES

To invert

$$\begin{bmatrix} 1 & -\frac{1}{2} & 0 \\ -\frac{1}{2} & 1 & -\frac{1}{2} \\ 0 & -\frac{1}{2} & 1 \end{bmatrix}$$

do:

1. Find the matrix of codeterminants (for each element, remove the corresponding row and column and find the determinant of what is left):

$$\begin{bmatrix} \frac{3}{4} & -\frac{1}{2} & \frac{1}{4} \\ -\frac{1}{2} & 1 & -\frac{1}{2} \\ \frac{1}{4} & -\frac{1}{2} & \frac{3}{4} \end{bmatrix}.$$

2. Change the sign of each element of the previous matrix according to the following checkerboard scheme:

$$\begin{bmatrix} + & - & + \\ - & + & - \\ + & - & + \end{bmatrix},$$

resulting in

$$
\begin{bmatrix}
\frac{3}{4} & \frac{1}{2} & \frac{1}{4} \\
\frac{1}{2} & 1 & \frac{1}{2} \\
\frac{1}{4} & \frac{1}{2} & \frac{3}{4}
\end{bmatrix}
$$

(all elements of $\mathbb{F}$ must be nonnegative).

3. Transpose the result:

$$
\begin{bmatrix}
\frac{3}{4} & \frac{1}{2} & \frac{1}{4} \\
\frac{1}{2} & 1 & \frac{1}{2} \\
\frac{1}{4} & \frac{1}{2} & \frac{3}{4}
\end{bmatrix}
$$

(nothing changes in this particular case, since the matrix was symmetric).

4. Divide each element by the determinant of the original matrix (found easily as the dot product of the first row of the original matrix and the first column of the previous matrix):

$$
\begin{bmatrix}
\frac{3}{2} & 1 & \frac{1}{2} \\
1 & 2 & 1 \\
\frac{1}{2} & 1 & \frac{3}{2}
\end{bmatrix}
$$

Remark 2.2. The number of operations required by this algorithm is proportional to $n!$ (n being the size of the matrix). This makes the algorithm practical for small matrices only (in our case, no more than 4×4) and *impossible* (even when using supercomputers) for matrices beyond even a moderate size (say 30×30).

INVERTING MATRICES (OF ANY SIZE)

The general procedure (easy to code) requires the following steps:

1. Append the unit matrix to the matrix to be inverted (creating a new matrix with twice as many columns as the old one), for example,

$$
\left[
\begin{array}{rrrr|rrrr}
2 & -3 & 5 & 1 & 1 & 0 & 0 & 0 \\
-1 & 4 & 0 & 5 & 0 & 1 & 0 & 0 \\
2 & -6 & 2 & 7 & 0 & 0 & 1 & 0 \\
1 & -3 & -4 & 3 & 0 & 0 & 0 & 1
\end{array}
\right].
$$

2. Use any number of the following ELEMENTARY OPERATIONS:

  - Multiply each element of a single row by the same nonzero constant;
  - Add/subtract a multiple of a row to/from any other row;
  - Interchange any two rows,

  to convert the original matrix to the unit matrix. Do this column by column: start with the main diagonal element (making it equal to 1), then make the remaining elements of the same column equal to 0.

3. The right side of the result (where the *original* unit matrix used to be) is the corresponding inverse.

If you fail to complete these steps (which can happen only when getting a *zero* on the *main diagonal and* every other element of the same column *below* the main diagonal), the original matrix is SINGULAR.

The number of operations required by this procedure is proportional to n^3 (n being the size of the matrix). In practical terms, this means even a standard laptop can invert matrices of huge size (say, $1,000^2$ elements) in a fraction of a second.

EXERCISES

Exercise 2.1. Consider a simple Markov chain with the following TPM:

$$\mathbb{P} = \begin{bmatrix} 0.2 & 0.3 & 0.5 \\ 0.1 & 0.5 & 0.4 \\ 0.6 & 0.2 & 0.2 \end{bmatrix}.$$

Assuming X_0 is generated from the distribution

X_0	1	2	3
Pr	0.6	0.0	0.4

find:

(a) $\Pr(X_2 = 3 \mid X_4 = 1)$;
(b) The stationary vector;
(c) The expected number of transitions it will take to enter, for the first time, State 2 and the corresponding standard deviation.

Exercise 2.2. Find (in terms of exact fractions) the fixed vector of the following TPM:

$$
\mathbb{P} =
\begin{bmatrix}
0 & 0.3 & 0.4 & 0 & 0.3 \\
0.4 & 0 & 0 & 0.6 & 0 \\
0.7 & 0 & 0 & 0.3 & 0 \\
0 & 0.5 & 0.1 & 0 & 0.4 \\
0.5 & 0 & 0 & 0.5 & 0
\end{bmatrix}
$$

and the limit $\lim\limits_{n \to \infty} \mathbb{P}^{2n}$.

(a) What is the long-run percentage of time spent in State 4?
(b) Is this Markov chain reversible (usually one can get the answer by constructing only a single element of $\overset{\circ}{\mathbb{P}}$)?

Exercise 2.3. Find the exact (in terms of fractions) answer to

$$
\lim_{n \to \infty}
\begin{bmatrix}
1 & 0 & 0 & 0 & 0 \\
0 & 0.15 & 0.85 & 0 & 0 \\
0 & 0.55 & 0.45 & 0 & 0 \\
0.12 & 0.18 & 0.21 & 0.26 & 0.23 \\
0.19 & 0.16 & 0.14 & 0.27 & 0.24
\end{bmatrix}^{n}.
$$

Exercise 2.4. Do the complete classification of the following TPM ($\times$ indicates a nonzero entry, $\cdot$ denotes zero):

$$
\begin{bmatrix}
\cdot & \cdot & \times & \times & \cdot & \cdot & \cdot \\
\cdot & \cdot & \times & \times & \cdot & \cdot & \cdot \\
\cdot & \cdot & \times & \cdot & \cdot & \cdot & \cdot \\
\cdot & \cdot & \times & \cdot & \cdot & \times & \times \\
\times & \cdot & \times & \cdot & \cdot & \cdot & \cdot \\
\cdot & \times & \times & \cdot & \cdot & \cdot & \cdot \\
\cdot & \cdot & \times & \cdot & \times & \cdot & \cdot
\end{bmatrix}
$$

Are any of the TPM's classes periodic?

Exercise 2.5. Using the TPM

$$\begin{bmatrix} 0.6 & 0.2 & 0.2 \\ 0.3 & 0.4 & 0.3 \\ 0.5 & 0.1 & 0.4 \end{bmatrix}$$

find $\Pr(X_3 = 2 \cap X_1 = 3)$ given that the initial state is drawn from the distribution

X_0	1	2	3
Pr	0.25	0.40	0.35

.

Also, find the probability of visiting State 2 before State 3.

Exercise 2.6. Find the fixed probability vector of the following TPM:

$$\begin{bmatrix} 0 & 0.4 & 0 & 0.2 & 0 & 0.4 \\ 0 & 0 & 0.7 & 0 & 0.3 & 0 \\ 1 & 0 & 0 & 0 & 0 & 0 \\ 0 & 0 & 0.4 & 0 & 0.6 & 0 \\ 1 & 0 & 0 & 0 & 0 & 0 \\ 0 & 0 & 0.2 & 0 & 0.8 & 0 \end{bmatrix}$$

.

Also, find (in exact fractions) $\lim_{n \to \infty} \mathbb{P}^{3n+1}$.

Exercise 2.7. Find the fixed probability vector of

$$\begin{bmatrix} 0 & 0.4 & 0 & 0.6 \\ 0.2 & 0 & 0.8 & 0 \\ 0 & 0.5 & 0 & 0.5 \\ 0.7 & 0 & 0.3 & 0 \end{bmatrix}$$

.

Starting in State 1, what is the probability of being in State 4 after 1,001 transitions?

Exercise 2.8. Calculate exactly using fractions:

$$\lim_{n\to\infty} \begin{bmatrix} 0.5 & 0.5 & 0 & 0 & 0 & 0 \\ 0.2 & 0.8 & 0 & 0 & 0 & 0 \\ 0 & 0 & 0 & 0 & 1 & 0 \\ 0 & 0 & 0 & 0 & 1 & 0 \\ 0 & 0 & 0.4 & 0.6 & 0 & 0 \\ 0.1 & 0 & 0 & 0.2 & 0.3 & 0.4 \end{bmatrix}^{2n+1}$$

Exercise 2.9. Do a complete classification of

$$\begin{bmatrix} \times & \cdot & \cdot & \cdot & \cdot & \cdot & \cdot & \times \\ \cdot & \cdot & \cdot & \cdot & \cdot & \times & \cdot & \cdot \\ \cdot & \cdot & \cdot & \cdot & \times & \cdot & \cdot & \times \\ \cdot & \cdot & \cdot & \cdot & \cdot & \cdot & \times & \cdot \\ \cdot & \cdot & \times & \cdot & \times & \cdot & \cdot & \cdot \\ \cdot & \times & \cdot & \cdot & \cdot & \cdot & \cdot & \cdot \\ \cdot & \cdot & \cdot & \times & \cdot & \cdot & \times & \cdot \\ \times & \cdot & \cdot & \cdot & \cdot & \times & \cdot & \cdot \end{bmatrix}$$

Exercise 2.10. Do a complete classification of

$$\begin{bmatrix} \times & \cdot & \cdot & \cdot & \times & \cdot & \times & \cdot & \cdot & \cdot \\ \cdot & \times & \cdot & \cdot & \cdot & \cdot & \cdot & \cdot & \cdot & \times \\ \cdot & \cdot & \cdot & \cdot & \times & \cdot & \cdot & \times & \cdot & \cdot \\ \cdot & \cdot & \cdot & \times & \cdot & \cdot & \cdot & \cdot & \times & \cdot \\ \times & \cdot & \cdot & \cdot & \cdot & \cdot & \cdot & \cdot & \cdot & \cdot \\ \cdot & \cdot & \cdot & \cdot & \cdot & \times & \cdot & \times & \cdot & \cdot \\ \times & \cdot & \cdot & \cdot & \cdot & \cdot & \times & \cdot & \cdot & \cdot \\ \cdot & \cdot & \times & \cdot & \cdot & \times & \cdot & \times & \cdot & \cdot \\ \cdot & \times & \cdot & \times & \cdot & \cdot & \cdot & \times & \cdot & \cdot \\ \cdot & \times & \cdot & \cdot & \cdot & \cdot & \cdot & \cdot & \cdot & \cdot \end{bmatrix}$$

Exercise 2.11. Do a complete classification of

$$
\begin{bmatrix}
\times & \cdot & \times & \cdot & \cdot & \times & \times & \cdot & \cdot \\
\cdot & \cdot & \cdot & \cdot & \times & \cdot & \cdot & \cdot & \cdot \\
\cdot & \cdot & \times & \cdot & \cdot & \cdot & \times & \cdot & \cdot \\
\times & \times & \cdot & \times & \times & \times & \cdot & \times & \times \\
\cdot & \times & \cdot & \cdot & \cdot & \cdot & \cdot & \cdot & \times \\
\times & \cdot & \times & \cdot & \cdot & \times & \times & \cdot & \cdot \\
\cdot & \cdot & \times & \cdot & \cdot & \cdot & \times & \cdot & \cdot \\
\times & \times & \cdot & \times & \times & \times & \cdot & \times & \times \\
\cdot & \cdot & \cdot & \cdot & \times & \cdot & \cdot & \cdot & \cdot
\end{bmatrix}.
$$

Exercise 2.12. Calculate exactly using fractions:

$$
\lim_{n \to \infty}
\begin{bmatrix}
0 & 0.5 & 0 & 0.2 & 0 & 0 & 0.3 & 0 \\
0 & 0' & 0.1 & 0 & 0.5 & 0.4 & 0 & 0 \\
0.2 & 0 & 0 & 0 & 0 & 0 & 0 & 0.8 \\
0 & 0 & 0.3 & 0 & 0.4 & 0.3 & 0 & 0 \\
0.7 & 0 & 0 & 0 & 0 & 0 & 0 & 0.3 \\
0.5 & 0 & 0 & 0 & 0 & 0 & 0 & 0.5 \\
0 & 0 & 0.2 & 0 & 0.6 & 0.2 & 0 & 0 \\
0 & 0.7 & 0 & 0.1 & 0 & 0 & 0.2 & 0
\end{bmatrix}^{3n+2}.
$$

Exercise 2.13. Calculate exactly using fractions:

$$
\lim_{n \to \infty}
\begin{bmatrix}
0 & 0 & 0.3 & 0.7 & 0 & 0 & 0 \\
0 & 0 & 0.2 & 0.8 & 0 & 0 & 0 \\
0 & 0 & 0 & 0 & 0.4 & 0.6 & 0 \\
0 & 0 & 0 & 0 & 0.5 & 0.5 & 0 \\
0.6 & 0.4 & 0 & 0 & 0 & 0 & 0 \\
0 & 1 & 0 & 0 & 0 & 0 & 0 \\
0.1 & 0.1 & 0.1 & 0.1 & 0.1 & 0.1 & 0.4
\end{bmatrix}^{3n+1}.
$$

Exercise 2.14. For

$$
\mathbb{P} =
\begin{bmatrix}
0.2 & 0 & 0.3 & 0.1 & 0 & 0.3 & 0.1 \\
0 & 0 & 0 & 0.7 & 0 & 0 & 0.3 \\
0.1 & 0.2 & 0.2 & 0 & 0 & 0.5 & 0 \\
0 & 0.7 & 0 & 0 & 0 & 0.3 & 0 \\
0 & 0.5 & 0 & 0 & 0 & 0.5 & 0 \\
0 & 0 & 0 & 0.2 & 0.2 & 0 & 0.6 \\
0 & 0.4 & 0 & 0 & 0 & 0.6 & 0
\end{bmatrix}
$$

find the exact (i.e., use fractions) value of $\lim_{n\to\infty} \mathbb{P}^{2n}$ and $\lim_{n\to\infty} \mathbb{P}^{2n+1}$.

CHAPTER 3
Finite Markov Chains II

We continue the study of finite Markov chains (FMCs) by considering models with one or more ABSORBING STATES. As their name implies, these are states that cannot be left once entered. Thus, a process entering an absorbing state is stuck there for good.

We also investigate the TIME REVERSAL of a Markov chain, that is, observing the process backward in time, and other special issues.

3.1 ABSORPTION OF TRANSIENT STATES

We have yet to figure out what happens to the transient-to-recurrent part (lower left corner) of $\mathbb{P}^n$ when n is large. To do this, we first must study the issue of the LUMPABILITY of states.

LUMPING OF STATES

Example 3.1. Let us return to our weather example with the following transition probability matrix (TPM):

$$\mathbb{P} = \begin{bmatrix} 0.6 & 0.3 & 0.1 \\ 0.4 & 0.5 & 0.1 \\ 0.3 & 0.4 & 0.3 \end{bmatrix}.$$

Is it possible to simplify the corresponding FMC by reducing the number of states (say by combining S and C into "fair" weather) without destroying the Markovian property?

J. Vrbik and P. Vrbik, *Informal Introduction to Stochastic Processes with Maple*, Universitext, DOI 10.1007/978-1-4614-4057-4_3,
© Springer Science+Business Media, LLC 2013

Solution. The answer is NO in general, YES if we are lucky and certain conditions are met. These amount to first partitioning the original matrix into the corresponding blocks:

	S	C	R
S	0.6	0.3	0.1
C	0.4	0.5	0.1
R	0.3	0.4	0.3

and then checking whether, in *each block*, the row sums are all *identical*. If so, the corresponding reduced matrix is a TPM of a new, simplified Markov chain. In our case we get

	F	R
F	0.9	0.1
R	0.7	0.3

On the other hand, if we keep sunny weather as such and classify C and R as "bad" weather, the results will no longer be consistent with an FMC model.

□

What follows are important examples of states that can always be lumped into a single state without destroying the Markovian property.

1. States of the same recurrent class—this we can do with as many recurrent classes as we like (possibly all—in which case the recurrent → recurrent part of $\mathbb{P}$ becomes the unit matrix).
2. Several (possibly all) recurrent classes can be lumped together into a single superstate.

(Note each of the new states defined so far will be *absorbing*.)

3. States of the same subclass (of a periodic class) can be lumped together (this is usually done with *each* subclass).

REDUCING RECURRENT CLASSES TO ABSORBING STATES

If we lump *each* of the recurrent classes into a single state, the new TPM (denoted by $\overline{\mathbb{P}}$) will acquire the following form:

$$\overline{\mathbb{P}} = \begin{bmatrix} \mathbb{I} & \mathbb{O} \\ \overline{\mathbb{U}} & \mathbb{T} \end{bmatrix},$$

where $\mathbb{I}$ is the unit matrix and $\overline{\mathbb{U}}$ is the corresponding lumped $\mathbb{U}$. By raising $\overline{\mathbb{P}}$ to the nth power, we get

$$\overline{\mathbb{P}}^n = \begin{bmatrix} \mathbb{I} & \mathbb{O} \\ \overline{\mathbb{U}} + \mathbb{T}\overline{\mathbb{U}} + \mathbb{T}^2\overline{\mathbb{U}} + \mathbb{T}^3\overline{\mathbb{U}} + \cdots + \mathbb{T}^{n-1}\overline{\mathbb{U}} & \mathbb{T}^n \end{bmatrix}. \tag{3.1}$$

In the $n \to \infty$ limit, we already know $\mathbb{T}^n \to \mathbb{O}$. The matrix in the lower left corner tends to $(\mathbb{I} + \mathbb{T} + \mathbb{T}^2 + \mathbb{T}^3 + \cdots)\overline{\mathbb{U}}$, where the first factor (the infinite sum of all $\mathbb{T}$-powers) can be computed by

$$\mathbb{F} \equiv \mathbb{I} + \mathbb{T} + \mathbb{T}^2 + \mathbb{T}^3 + \cdots = (\mathbb{I} - \mathbb{T})^{-1}.$$

This is a natural (and legitimate) extension of the usual geometric-summation formula, with the reciprocal replaced by the *matrix inverse*. $\mathbb{F}$ is called the FUNDAMENTAL MATRIX of the corresponding FMC (note it is defined based solely on the transient → transient part of $\mathbb{P}$). With the help of $\mathbb{F}$, the limit of $\overline{\mathbb{P}}^n$ as $n \to \infty$ ($\overline{\mathbb{P}}^\infty$ for short) can be written as

$$\overline{\mathbb{P}}^\infty = \begin{bmatrix} \mathbb{I} & \mathbb{O} \\ \mathbb{F}\overline{\mathbb{U}} & \mathbb{O} \end{bmatrix},$$

where the elements of the lower-left corner of this matrix represent probabilities of being absorbed (sooner or later) in the corresponding recurrent class (column) for each of the transient states (row).

Example 3.2. Returning to our betting example, we can rearrange the PT-M as follows:

$$
\begin{array}{c|ccccc}
 & -2 & 2 & -1 & 0 & 1 \\
\hline
-2 & 1 & 0 & 0 & 0 & 0 \\
2 & 0 & 1 & 0 & 0 & 0 \\
\hline
-1 & \frac{1}{2} & 0 & 0 & \frac{1}{2} & 0 \\
0 & 0 & 0 & \frac{1}{2} & 0 & \frac{1}{2} \\
1 & 0 & \frac{1}{2} & 0 & \frac{1}{2} & 0 \\
\end{array}
\tag{3.2}
$$

Because both of our recurrent states are absorbing, there is no need to construct $\overline{\mathbb{P}}$ ($\mathbb{P}$ itself has the required structure). By computing

$$\mathbb{F} = \begin{bmatrix} 1 & -\frac{1}{2} & 0 \\ -\frac{1}{2} & 1 & -\frac{1}{2} \\ 0 & -\frac{1}{2} & 1 \end{bmatrix}^{-1} = \begin{bmatrix} \frac{3}{2} & 1 & \frac{1}{2} \\ 1 & 2 & 1 \\ \frac{1}{2} & 1 & \frac{3}{2} \end{bmatrix}$$

and

$$\mathbb{F} \begin{bmatrix} \frac{1}{2} & 0 \\ 0 & 0 \\ 0 & \frac{1}{2} \end{bmatrix} = \begin{bmatrix} \frac{3}{4} & \frac{1}{4} \\ \frac{1}{2} & \frac{1}{2} \\ \frac{1}{4} & \frac{3}{4} \end{bmatrix}$$

we get the individual probabilities of absorption, that is, winning or losing the game given we start with \$3 (first row), \$2 (second row), or \$1 (third row). $\diamondsuit$

We should mention that if we organize the transient states into classes and arrange these from "top to bottom" (so that transitions can go only from a higher to a lower class), then the $\mathbb{I} - \mathbb{T}$ matrix becomes block upper triangular and easier to invert by utilizing

$$\begin{bmatrix} \mathbb{A} & \mathbb{B} \\ \mathbb{O} & \mathbb{C} \end{bmatrix}^{-1} = \begin{bmatrix} \mathbb{A}^{-1} & \mathbb{A}^{-1}\mathbb{B}\mathbb{C}^{-1} \\ \mathbb{O} & \mathbb{C}^{-1} \end{bmatrix}.$$

Note only the main-diagonal blocks need to be inverted (it is a lot easier to invert two 2×2 matrices than one 4×4 matrix). This is important mainly when we are forced to perform all computations by hand (we review the procedure for inverting matrices in Appendix 2.A).

TIME UNTIL ABSORPTION

It is also interesting to investigate the number of transitions (a random variable, say Y) necessary to reach any of the absorbing states (in the present case, this represents the game's duration). For this purpose, we can lump all recurrent states into a single superstate, leading to

$$\overline{\overline{\mathbb{P}}} = \begin{bmatrix} 1 & \mathbb{O} \\ \overline{\overline{\mathbb{U}}} & \mathbb{T} \end{bmatrix},$$

where $\overline{\overline{\mathbb{U}}}$ has only one column. The individual powers of $\overline{\overline{\mathbb{P}}}$ [which we know how to compute – see (3.1)] yield, in the lower left corner, the probability of having been already absorbed (having finished the game), after taking that many (powers of $\overline{\overline{\mathbb{P}}}$) transitions. The differences then yield the probability of absorption *during* that particular transition, namely:

Y (transitions to absorption)	1	2	3	4	$\cdots$
Pr	$\overline{\overline{\mathbb{U}}}$	$\mathbb{T}\overline{\overline{\mathbb{U}}}$	$\mathbb{T}^2\overline{\overline{\mathbb{U}}}$	$\mathbb{T}^3\overline{\overline{\mathbb{U}}}$	$\cdots$

The corresponding expected value, say $\tau \equiv \mathbb{E}(Y)$, is given by

$$(\mathbb{I} + 2\mathbb{T} + 3\mathbb{T}^2 + 4\mathbb{T}^3 + \cdots)\overline{\overline{\mathbb{U}}} = \mathbb{F}^2\overline{\overline{\mathbb{U}}},$$

analogously to $1 + 2x + 3x^2 + 4x^3 + \cdots = (1 + x + x^2 + x^3 + x^4 + \cdots)' = \left(\frac{1}{1-x}\right)' = \frac{1}{(1-x)^2}$.

Since $\overline{\overline{\mathbb{U}}}$ and $\mathbb{F}$ are closely related, we can actually simplify the preceding formula using the following proposition.

Proposition 3.1. Let $\mathrm{Sum}(\mathbb{A})$ be the column vector of row sums of $\mathbb{A}$, that is, the vector whose ith row/entry is given by $\sum_j \mathbb{A}_{ij}$. Then

$$\mathbb{A}\,\mathrm{Sum}(\mathbb{B}) = \mathrm{Sum}(\mathbb{A}\mathbb{B}),$$

where $\mathbb{A}$ and $\mathbb{B}$ are two compatible matrices.

Proof. We have

$$\sum_j \mathbb{A}_{ij} \sum_k \mathbb{B}_{jk} = \sum_k \sum_j \mathbb{A}_{ij}\mathbb{B}_{jk}.$$

$\square$

This implies

$$\begin{aligned}
\tau &= \mathbb{F}^2\overline{\overline{\mathbb{U}}} \\
&= (\mathbb{I} - \mathbb{T})^{-2}\mathrm{Sum}(\mathbb{I} - \mathbb{T}) \\
&= \mathrm{Sum}\left((\mathbb{I} - \mathbb{T})^{-2}(\mathbb{I} - \mathbb{T})\right) \\
&= \mathrm{Sum}\left((\mathbb{I} - \mathbb{T})^{-1}\right) \\
&= \mathrm{Sum}(\mathbb{F}).
\end{aligned}$$

Proposition 3.2. τ *can also be computed as the unique solution to*

$$(\mathbb{I} - \mathbb{T})\tau = \begin{bmatrix} 1 \\ 1 \\ \vdots \\ 1 \end{bmatrix}.$$

Proof. Take Sum of each side of $(\mathbb{I} - \mathbb{T})\mathbb{F} = \mathbb{I}$. $\square$

Using our previous example, this results in $\tau = [3, 4, 3]^{\mathrm{T}}$ for the expected number of rounds of the game.

Note $\mathbb{F}$ itself (since $\mathbb{T}^n$ yields the probability of being in a specific transient state after n transitions, given the initial state) represents the expected number of visits to each transient state, given the initial state – being in this state *initially* counts as one visit, which is why the diagonal elements of $\mathbb{F}$ must always be greater than 1.

Similarly, since

$$2 \times 1 + 3 \times 2x + 4 \times 3x^2 + 5 \times 4x^3 + \cdots = \left(\frac{1}{1-x}\right)'' = \frac{2}{(1-x)^3},$$

we get

$$\mathbb{E}\left((Y+1)Y\right) = 2\mathbb{F}^3\overline{\overline{\mathbb{U}}} = 2\mathbb{F}^3\mathrm{Sum}(\mathbb{I} - \mathbb{T}) = 2\mathrm{Sum}(\mathbb{F}^2) = 2\mathbb{F}\mathrm{Sum}(\mathbb{F}) = 2\mathbb{F}\boldsymbol{\tau}.$$

This implies

$$\mathrm{Var}(Y) = 2\mathbb{F}\boldsymbol{\tau} - \boldsymbol{\tau} - \boldsymbol{\tau}^{sq},$$

where $\boldsymbol{\tau}^{sq}$ represents *componentwise* squaring of the elements of $\boldsymbol{\tau}$.

For our "betting" example, this yields

$$2 \begin{bmatrix} \frac{3}{2} & 1 & \frac{1}{2} \\ 1 & 2 & 1 \\ \frac{1}{2} & 1 & \frac{3}{2} \end{bmatrix} \begin{bmatrix} 3 \\ 4 \\ 3 \end{bmatrix} - \begin{bmatrix} 3 \\ 4 \\ 3 \end{bmatrix} - \begin{bmatrix} 9 \\ 16 \\ 9 \end{bmatrix} = \begin{bmatrix} 8 \\ 8 \\ 8 \end{bmatrix}.$$

INITIAL DISTRIBUTION

When the initial state is generated using a probability distribution $\mathbf{d}$, one gets

$$\Pr(Y = n) = \sum_{i=1}^{N} d_i \Pr(Y = n \mid X_0 = i).$$

The expected value of Y is thus

$$\mathbb{E}(Y) = \sum_{n=1}^{\infty} n \Pr(Y = n) = \sum_{i=1}^{N} d_i \sum_{n=1}^{\infty} n \Pr(Y = n \mid X_0 = i) = \sum_{i=1}^{N} d_i \tau_i = \mathbf{d}^{\mathrm{T}}\boldsymbol{\tau}.$$

Similarly,

$$\mathbb{E}(Y^2) = \sum_{n=1}^{\infty} n^2 \Pr(Y = n)$$

$$= \sum_{i=1}^{N} d_i \sum_{n=1}^{\infty} n^2 \Pr(Y = n \mid X_0 = i)$$

$$= \sum_{i=1}^{N} d_i \mathbb{E}(Y^2 \mid X_0 = i)$$

$$= \sum_{i=1}^{N} d_i (2\mathbb{F}\boldsymbol{\tau} - \boldsymbol{\tau})_i$$

$$= \mathbf{d}^{\mathrm{T}} (2\mathbb{F}\boldsymbol{\tau} - \boldsymbol{\tau}).$$

But note $\mathrm{Var}(Y)$ *does not equal* $\sum_{i=1}^{N} d_i \mathrm{Var}(Y \mid X_0 = i)$! Instead, we have

$$\mathrm{Var}(Y) = \sum_{i=1}^{N} d_i (2\mathbb{F}\boldsymbol{\tau} - \boldsymbol{\tau})_i - \left(\sum_{i=1}^{N} d_i \tau_i \right)^2.$$

Example 3.3. If we flip a coin repeatedly, what is the probability of generating the TTT pattern *before* HH?

Solution. We must first consider *three* consecutive symbols as a single state of an FMC, resulting in the following TPM:

	HHH	HHT	HTH	HTT	THH	THT	TTH	TTT	
HHH	$\frac{1}{2}$	$\frac{1}{2}$	$\cdot$	$\cdot$	$\cdot$	$\cdot$	$\cdot$	$\cdot$	
HHT	$\cdot$	$\cdot$	$\frac{1}{2}$	$\frac{1}{2}$	$\cdot$	$\cdot$	$\cdot$	$\cdot$	
HTH	$\cdot$	$\cdot$	$\cdot$	$\cdot$	$\frac{1}{2}$	$\frac{1}{2}$	$\cdot$	$\cdot$	
HTT	$\cdot$	$\cdot$	$\cdot$	$\cdot$	$\cdot$	$\cdot$	$\frac{1}{2}$	$\frac{1}{2}$	(3.3)
THH	$\frac{1}{2}$	$\frac{1}{2}$	$\cdot$	$\cdot$	$\cdot$	$\cdot$	$\cdot$	$\cdot$	
THT	$\cdot$	$\cdot$	$\frac{1}{2}$	$\frac{1}{2}$	$\cdot$	$\cdot$	$\cdot$	$\cdot$	
TTH	$\cdot$	$\cdot$	$\cdot$	$\cdot$	$\frac{1}{2}$	$\frac{1}{2}$	$\cdot$	$\cdot$	
TTT	$\cdot$	$\cdot$	$\cdot$	$\cdot$	$\cdot$	$\cdot$	$\frac{1}{2}$	$\frac{1}{2}$	

We then make TTT, HHH, HHT, and THH absorbing, lumping the last three together as a single state HH, and thus

	HH	HTH	HTT	THT	TTH	TTT
HH	1	0	0	0	0	0
HTH	$\frac{1}{2}$	0	0	$\frac{1}{2}$	0	0
HTT	0	0	0	0	$\frac{1}{2}$	$\frac{1}{2}$
THT	0	$\frac{1}{2}$	$\frac{1}{2}$	0	0	0
TTH	$\frac{1}{2}$	0	0	$\frac{1}{2}$	0	0
TTT	0	0	0	0	0	1

The fundamental matrix is thus equal to

$$
\mathbb{F} =
\begin{bmatrix}
1 & 0 & -\frac{1}{2} & 0 \\
0 & 1 & 0 & -\frac{1}{2} \\
-\frac{1}{2} & -\frac{1}{2} & 1 & 0 \\
0 & 0 & -\frac{1}{2} & 1
\end{bmatrix}^{-1}
=
\begin{bmatrix}
1.4 & 0.4 & 0.8 & 0.2 \\
0.2 & 1.2 & 0.4 & 0.6 \\
0.8 & 0.8 & 1.6 & 0.4 \\
0.4 & 0.4 & 0.8 & 1.2
\end{bmatrix},
$$

which implies

$$
\mathbb{F}\mathbb{U} =
\begin{bmatrix}
1.4 & 0.4 & 0.8 & 0.2 \\
0.2 & 1.2 & 0.4 & 0.6 \\
0.8 & 0.8 & 1.6 & 0.4 \\
0.4 & 0.4 & 0.8 & 1.2
\end{bmatrix}
\begin{bmatrix}
\frac{1}{2} & 0 \\
0 & \frac{1}{2} \\
0 & 0 \\
\frac{1}{2} & 0
\end{bmatrix}
=
\begin{bmatrix}
0.8 & 0.2 \\
0.4 & 0.6 \\
0.6 & 0.4 \\
0.8 & 0.2
\end{bmatrix}.
$$

This can be expanded to cover all of the original eight states; thus,

$$
\begin{bmatrix}
1 & 0 \\
1 & 0 \\
0.8 & 0.2 \\
0.4 & 0.6 \\
1 & 0 \\
0.6 & 0.4 \\
0.8 & 0.2 \\
0 & 1
\end{bmatrix}
$$

(the first/second column giving the probabilities of "being absorbed" by HH/TTT). Since the initial probabilities are equal for all eight states, we must simply average the second column to get the final answer: TTT wins over HH with a probability of $\frac{2.4}{8} = 30\%$.

$\Diamond$

Example 3.4. (Continuation of Example 3.3). What is the expected duration (in terms of number of flips) and the corresponding standard deviation of this game?

Solution. Based on $\mathbb{F}$,

$$
\tau_Y = [2.8, 2.4, 3.6, 2.8]^{\mathrm{T}},
$$

that is, the expected number of transitions (or flips) *after* the initial state has been generated. To count all the *flips* required to finish the game (let us call this random variable V), we must add 3 to Y (and therefore to each of the preceding expected values). Furthermore, we can extend τ_Y to cover all possible states (not just the transients); thus,

$$\tau = [2, 2, 5.8, 5.4, 3, 6.6, 5.8, 3]^T$$

(note HHH and HHT would result in ending the game in two flips, not three). Since each of the eight initial states has the same probability of being generated, the ordinary average of elements of τ_V, namely, $\frac{33.6}{8} = 4.2$, is the expected number of flips to conclude this game.

A similar approach enables us to evaluate $\mathbb{E}(V^2)$. First we compute

$$\mathbb{E}(Y^2 \mid X_0 = i) = 2\mathbb{F}\tau_Y - \tau_Y = \begin{bmatrix} 13.84 \\ 10.72 \\ 18.48 \\ 13.84 \end{bmatrix}.$$

This can be easily extended to $\mathbb{E}(V^2)$ [equal to $\mathbb{E}(Y^2) + 6\mathbb{E}(Y) + 9$ for the transient states, to 2^2 for HHH and HHT, and to 3^2 for THH and TTT]:

$$\mathbb{E}(V^2 \mid X_0 = i) = \begin{bmatrix} 4 \\ 4 \\ 39.64 \\ 34.12 \\ 9 \\ 49.08 \\ 39.64 \\ 9 \end{bmatrix}.$$

The corresponding average is $\frac{188.48}{8} = 23.56$. The variance of V is thus equal to $23.56 - 4.2^2 = 5.92$, and the corresponding standard deviation is 2.433. ◻

Example 3.5. [A further extension of Example 3.3] What is the probability of finishing this game without ever visiting THT?

Solution. What we must do now is to make THT absorbing as well. The question then is simply: Do we get absorbed in THT or in one of our competing states HH and TTT (which, for the purpose of this question, can be further lumped into a single game-over state)? The corresponding TPM will look like this:

	GO	THT	HTH	HTT	TTH
GO	1	0	0	0	0
THT	0	1	0	0	0
HTH	$\frac{1}{2}$	$\frac{1}{2}$	0	0	0
HTT	$\frac{1}{2}$	0	0	0	$\frac{1}{2}$
TTH	$\frac{1}{2}$	$\frac{1}{2}$	0	0	0

which implies

$$\mathbb{FU} = \begin{bmatrix} 1 & 0 & 0 \\ 0 & 1 & -\frac{1}{2} \\ 0 & 0 & 1 \end{bmatrix}^{-1} \begin{bmatrix} \frac{1}{2} & \frac{1}{2} \\ \frac{1}{2} & 0 \\ \frac{1}{2} & \frac{1}{2} \end{bmatrix} = \begin{bmatrix} \frac{1}{2} & \frac{1}{2} \\ \frac{3}{4} & \frac{1}{4} \\ \frac{1}{2} & \frac{1}{2} \end{bmatrix}$$

since

$$\mathbb{F} = \begin{bmatrix} 1 & 0 & 0 \\ 0 & 1 & \frac{1}{2} \\ 0 & 0 & 1 \end{bmatrix}.$$

The probability that the game will be over without visiting THT is thus $[\frac{1}{2}, \frac{3}{4}, \frac{1}{2}]^{\text{T}}$, given that we start in the corresponding transient state. This vector can be extended to cover all possible initial states:

$$\left[1, 1, \frac{1}{2}, \frac{3}{4}, 1, 0, \frac{1}{2}, 1\right]^{\text{T}}.$$

The average of these, namely, $\frac{5.75}{8} = 71.875\%$, yields the final answer. ⬦

   This is how we would deal in general with the question of being absorbed without ever visiting a specific transient state (or a collection of such states) – we make all these states absorbing as well!

   Similarly, we compute the so-called TABOO PROBABILITIES of a *regular* FMC: starting in State a, what is the probability of visiting State b before State c (make both b and c absorbing). Note, in the regular case, the probability of visiting b (and also of visiting c) sooner or later is 1, and the issue is: which state is reached *first*?

Example 3.6. Related to the last example is the following question: flipping a coin repeatedly, how often do we generate a certain pattern (say TTT)? This is a bit ambiguous: once the pattern is generated, do we allow any part of it to be the start of the next occurrence, or do we have to build the pattern from scratch (i.e., do we consider HTTTTTT as four occurrences of TTT, or only two)? If we allow the patterns to overlap, the issue is quite simple (and the answer is, in this case, 8 – why?), if we have to build them from scratch, we must make the pattern absorbing to come up with the right answer.

Solution. By making TTT absorbing in (3.3) we can get the corresponding τ_Y by solving

$$
\begin{bmatrix}
\frac{1}{2} & -\frac{1}{2} & 0 & 0 & 0 & 0 & 0 \\
0 & 1 & -\frac{1}{2} & -\frac{1}{2} & 0 & 0 & 0 \\
0 & 0 & 1 & 0 & -\frac{1}{2} & -\frac{1}{2} & 0 \\
0 & 0 & 0 & 1 & 0 & 0 & -\frac{1}{2} \\
-\frac{1}{2} & -\frac{1}{2} & 0 & 0 & 1 & 0 & 0 \\
0 & 0 & -\frac{1}{2} & -\frac{1}{2} & 0 & 1 & 0 \\
0 & 0 & 0 & 0 & -\frac{1}{2} & -\frac{1}{2} & 1
\end{bmatrix}
\tau_Y =
\begin{bmatrix}
1 \\ 1 \\ 1 \\ 1 \\ 1 \\ 1 \\ 1
\end{bmatrix}.
$$

The solution is

$$\tau_Y = [14,\ 12,\ 14,\ 8,\ 14,\ 12,\ 14]^{\mathrm{T}},$$

which can be verified by typing in MAPLE:

$$
> \mathbb{P} :=
\begin{bmatrix}
\frac{1}{2} & -\frac{1}{2} & 0 & 0 & 0 & 0 & 0 \\
0 & 1 & -\frac{1}{2} & -\frac{1}{2} & 0 & 0 & 0 \\
0 & 0 & 1 & 0 & -\frac{1}{2} & -\frac{1}{2} & 0 \\
0 & 0 & 0 & 1 & 0 & 0 & -\frac{1}{2} \\
-\frac{1}{2} & -\frac{1}{2} & 0 & 0 & 1 & 0 & 0 \\
0 & 0 & -\frac{1}{2} & -\frac{1}{2} & 0 & 1 & 0 \\
0 & 0 & 0 & 0 & -\frac{1}{2} & -\frac{1}{2} & 1
\end{bmatrix} :
$$

```
> τ := LinearSolve (ℙ, Vector (1..7, 1)) :
> convert (τ, list);
```
{We do this now, and later, only to save vertical space.}

$$[14,\ 12,\ 14,\ 8,\ 14,\ 12,\ 14]$$

Extended by an extra component equal to 0 (to include TTT itself) and then averaging over all initial possibilities results in $\frac{88}{8} = 11$ (this is the number of flips *after* the initial state has been generated). The TTT pattern is thus generated (from scratch), on average, every 14 flips $(11 + 3)$.

Similarly, $\mathbb{F}\boldsymbol{\tau}_Y$ can be found as the (unique) solution to

$$(\mathbb{I} - \mathbb{T})(\mathbb{F}\boldsymbol{\tau}_Y) = \boldsymbol{\tau}_Y,$$

namely,

$$\mathbb{F}\boldsymbol{\tau}_Y = [176, 148, 176, 96, 176, 148, 176]^{\mathrm{T}},$$

or, by MAPLE:

> *convert* $(LinearSolve(\mathbb{P}, \tau), list)$;

$$[176, 148, 176, 96, 176, 148, 176]$$

We thus get

$$\mathbb{E}(Y^2 \mid X_0 = i) = 2\mathbb{F}\boldsymbol{\tau}_Y - \boldsymbol{\tau}_Y = [338, 284, 338, 184, 338, 284, 338]^{\mathrm{T}}.$$

When extended by $\mathbb{E}(Y^2 \mid X_0 = \text{TTT}) = 0$ and averaged, this yields $\mathbb{E}(Y^2) = \frac{2104}{8} = 263$, implying $\text{Var}(Y) = 263 - 11^2 = 142$. Since $V \equiv Y + 3$, V has the same variance as Y. The corresponding standard deviation is then 11.92 (almost as large as the expected value itself). $\lozenge$

We will discuss a more elegant way of dealing with pattern generation of this type in the next chapter.

LARGE POWERS OF A STOCHASTIC MATRIX

We now return to our main issue of computing any large power of a TPM. It is quite easy to complete the task (including the lower left corner), provided all recurrent classes are *regular*. We already know how to construct $\overline{\mathbb{P}}^\infty$; the only question is how to "unlump" it. We also know the full form of the upper left corner of $\mathbb{P}^\infty$; we just need to figure out what happens (in the long run) when we get absorbed in a specific recurrent class. The answer is easy to guess: after many transitions, the probability the process will be in any of its individual states should follow the corresponding stationary distribution s. And this is indeed the case.

Example 3.7. Find

$$
\begin{bmatrix}
1 & 1 & 0 & 0 & 0 & 0 \\
0 & \frac{3}{4} & \frac{1}{4} & 0 & 0 \\
0 & \frac{1}{8} & \frac{7}{8} & 0 & 0 \\
\frac{1}{4} & \frac{1}{4} & 0 & \frac{1}{8} & \frac{3}{8} \\
\frac{1}{3} & 0 & \frac{1}{6} & \frac{1}{6} & \frac{1}{3}
\end{bmatrix}^{1000},
$$

where the recurrent classes are boxed (verify this classification).   *Solution.*

First we compute

$$
\overline{\mathbb{P}} =
\begin{bmatrix}
1 & 0 & 0 & 0 \\
0 & 1 & 0 & 0 \\
\frac{1}{4} & \frac{1}{4} & \frac{1}{8} & \frac{3}{8} \\
\frac{1}{3} & \frac{1}{6} & \frac{1}{6} & \frac{1}{3}
\end{bmatrix}
$$

then

$$
\overline{\mathbb{F}}\overline{\mathbb{U}} =
\begin{bmatrix}
\frac{7}{8} & -\frac{3}{8} \\
-\frac{1}{6} & \frac{2}{3}
\end{bmatrix}^{-1}
\begin{bmatrix}
\frac{1}{4} & \frac{1}{4} \\
\frac{1}{3} & \frac{1}{6}
\end{bmatrix}
=
\begin{bmatrix}
\frac{32}{25} & \frac{18}{25} \\
\frac{8}{25} & \frac{42}{25}
\end{bmatrix}
\begin{bmatrix}
\frac{1}{4} & \frac{1}{4} \\
\frac{1}{3} & \frac{1}{6}
\end{bmatrix}
=
\begin{bmatrix}
\frac{14}{25} & \frac{11}{25} \\
\frac{16}{25} & \frac{9}{25}
\end{bmatrix}.
$$

This implies (to a good accuracy)

$$
\overline{\mathbb{P}}^{1000} \approx
\begin{bmatrix}
1 & 0 & 0 & 0 \\
0 & 1 & 0 & 0 \\
\frac{14}{25} & \frac{11}{25} & 0 & 0 \\
\frac{16}{25} & \frac{9}{25} & 0 & 0
\end{bmatrix}.
$$

To expand this result to the full $\mathbb{P}^{1000}$, we must find the stationary probabilities of the second recurrent class. Solving

$$
\begin{bmatrix}
\frac{1}{4} & -\frac{1}{8} \\
-\frac{1}{4} & \frac{1}{8}
\end{bmatrix}
\mathbf{s} = \mathbf{0}
$$

we get $\mathbf{s} = \begin{bmatrix} \frac{1}{3} \\ \frac{2}{3} \end{bmatrix}$. All we have to do now is expand the second row (by simply duplicating it) and the second column of $\overline{\mathbb{P}}^{1000}$ (according to this $\mathbf{s}$); thus,

$$
\mathbb{P}^{1000} \approx \begin{bmatrix}
1 & 0 & 0 & 0 & 0 \\
0 & \frac{1}{3} & \frac{2}{3} & 0 & 0 \\
0 & \frac{1}{3} & \frac{2}{3} & 0 & 0 \\
\frac{14}{25} & \frac{11}{25} \times \frac{1}{3} & \frac{11}{25} \times \frac{2}{3} & 0 & 0 \\
\frac{16}{25} & \frac{9}{25} \times \frac{1}{3} & \frac{9}{25} \times \frac{2}{3} & 0 & 0
\end{bmatrix}.
$$

$\square$

PERIODIC CASE

When one (or more) of the recurrent classes are periodic, constructing large powers of $\mathbb{P}$ becomes a bit more difficult (to the extent that we normally deal with only one periodic class at a time).

We will thus assume there is only one recurrent class of period 3 (the general pattern will be clear from this example). If there are other recurrent classes, they can be all lumped into a single superstate and dealt with separately later on. We thus have

$$
\mathbb{P} = \begin{bmatrix}
\mathbb{O} & \mathbf{C}_1 & \mathbb{O} & \mathbb{O} \\
\mathbb{O} & \mathbb{O} & \mathbf{C}_2 & \mathbb{O} \\
\mathbf{C}_3 & \mathbb{O} & \mathbb{O} & \mathbb{O} \\
\mathbf{U}_1 & \mathbf{U}_2 & \mathbf{U}_3 & \mathbf{T}
\end{bmatrix}.
$$

By lumping the states of each subclass, we reduce this to

$$
\widetilde{\mathbb{P}} = \begin{bmatrix}
\mathbf{J} & \mathbb{O} \\
\widetilde{\mathbf{U}} & \mathbf{T}
\end{bmatrix},
$$

where

$$
\mathbf{J} = \begin{bmatrix}
0 & 1 & 0 \\
0 & 0 & 1 \\
1 & 0 & 0
\end{bmatrix}
$$

(a unit matrix, with each column moved, cyclically, one space to the right). Raising $\widetilde{\mathbb{P}}$ to the power of 3 (λ in general) yields

$$\widehat{\mathbb{P}} \equiv \widetilde{\mathbb{P}}^3 = \begin{bmatrix} \mathbb{I} & \mathbb{O} \\ \widetilde{\mathbb{U}}\mathbb{J}^2 + \mathbb{T}\widetilde{\mathbb{U}}\mathbb{J} + \mathbb{T}^2\widetilde{\mathbb{U}} & \mathbb{T}^3 \end{bmatrix} \equiv \begin{bmatrix} \mathbb{I} & \mathbb{O} \\ \widehat{\mathbb{U}} & \widehat{\mathbb{T}} \end{bmatrix}$$

(note right-multiplying a matrix by $\mathbb{J}$ cyclically moves each of its *columns* to the right – a fairly simple operation). And we already know $\widehat{\mathbb{P}}^\infty$ exists, and we know how to construct it – the bottom left corner is $(\mathbb{I} - \widehat{\mathbb{T}})^{-1}\widehat{\mathbb{U}}$. Furthermore, we know how to expand it (using the stationary probabilities of each subclass) to the full $\mathbb{P}^{3\infty}$ (this notation denotes $\lim_{n\to\infty} \mathbb{P}^{3n}$).

To get the answer for $\mathbb{P}^{3n+1}$ (n large), we first multiply $\widehat{\mathbb{P}}^\infty$ by $\widetilde{\mathbb{P}}$ (which cyclically rotates its first three columns to the right) and then expand each of these columns by the corresponding stationary probabilities. And, similarly, we get $\mathbb{P}^{3\infty+2}$ (cyclically rotate the first three columns of $\widehat{\mathbb{P}}^\infty$ by two positions to the right, and expand).

Example 3.8. Find

$$\begin{bmatrix} 0.3 & 0.7 & \cdot & \cdot & \cdot & \cdot & \cdot & \cdot \\ 0.6 & 0.4 & \cdot & \cdot & \cdot & \cdot & \cdot & \cdot \\ \cdot & \cdot & \cdot & \cdot & 0.2 & 0.8 & \cdot & \cdot \\ \cdot & \cdot & \cdot & \cdot & 0.5 & 0.5 & \cdot & \cdot \\ \cdot & \cdot & 0.7 & 0.3 & \cdot & \cdot & \cdot & \cdot \\ \cdot & \cdot & 0.9 & 0.1 & \cdot & \cdot & \cdot & \cdot \\ 0.1 & 0.1 & 0.1 & 0.1 & 0.2 & 0.1 & 0.1 & 0.2 \\ 0.1 & 0.2 & 0.2 & 0.1 & 0.1 & 0.1 & 0.1 & 0.1 \end{bmatrix}^{1000}$$

Solution. A simple classification will confirm States 1 and 2 constitute a regular recurrent class, $\{3, 4, 5, 6\}$ another recurrent class with a period of 2 (subclasses $\{3, 4\}$ and $\{5, 6\}$), and that States 7 and 8 are transient.

We will first take care of the regular class (and the corresponding columns):

$$\mathbb{F} \equiv (\mathbb{I} - \mathbb{T})^{-1} = \begin{bmatrix} \frac{90}{79} & \frac{20}{79} \\ \frac{10}{79} & \frac{90}{79} \end{bmatrix}$$

and

$$\overline{\mathbb{U}} = \begin{bmatrix} 0.2 & 0.5 \\ 0.3 & 0.5 \end{bmatrix} \Rightarrow \mathbb{F}\overline{\mathbb{U}} = \begin{bmatrix} \frac{24}{79} & \frac{55}{79} \\ \frac{29}{79} & \frac{50}{79} \end{bmatrix}.$$

The stationary probability vector is a solution to $(\mathbb{I} - \mathbb{P}^{\mathrm{T}})\mathbf{s} = \mathbf{0}$ or, more explicitly, to

$$\begin{bmatrix} 0.7 & -0.6 \\ -0.7 & 0.6 \end{bmatrix} \mathbf{s} = \mathbf{0}.$$

Triviailly, $\mathbf{s} = \begin{bmatrix} \frac{6}{13} \\ \frac{7}{13} \end{bmatrix}$. The first two columns of $\mathbb{P}^{1000}$ are thus (respectively) equal to

$$\begin{bmatrix} \frac{6}{13} & \frac{6}{13} & 0 & 0 & 0 & 0 & \frac{24}{79} \times \frac{6}{13} & \frac{29}{79} \times \frac{6}{13} \end{bmatrix}^{\mathrm{T}}$$

and

$$\begin{bmatrix} \frac{7}{13} & \frac{7}{13} & 0 & 0 & 0 & 0 & \frac{24}{79} \times \frac{7}{13} & \frac{29}{79} \times \frac{7}{13} \end{bmatrix}^{\mathrm{T}}.$$

To deal with the periodic-class columns (3, 4, 5, and 6) of $\mathbb{P}^{1000}$, we first find

$$\widetilde{\mathbb{P}} = \begin{bmatrix} 1 & \cdot & \cdot & \cdot & \cdot \\ \cdot & \cdot & 1 & \cdot & \cdot \\ \cdot & 1 & \cdot & \cdot & \cdot \\ 0.2 & 0.2 & 0.3 & 0.1 & 0.2 \\ 0.3 & 0.3 & 0.2 & 0.1 & 0.1 \end{bmatrix}.$$

Squaring this matrix yields

$$\widehat{\mathbb{P}} = \begin{bmatrix} 1 & \cdot & \cdot & \cdot & \cdot \\ \cdot & 1 & \cdot & \cdot & \cdot \\ \cdot & \cdot & 1 & \cdot & \cdot \\ 0.28 & 0.38 & 0.27 & 0.03 & 0.04 \\ 0.35 & 0.25 & 0.35 & 0.02 & 0.03 \end{bmatrix}$$

with its own

$$\widehat{\mathbb{F}} \equiv (\mathbb{I} - \widehat{\mathbb{T}})^{-1} = \begin{bmatrix} \frac{9700}{9401} & \frac{400}{9401} \\ \frac{200}{9401} & \frac{9700}{9401} \end{bmatrix}.$$

This, postmultiplied by $\widehat{U}$, yields

$$\begin{bmatrix} \frac{2856}{9401} & \frac{3786}{9401} & \frac{2759}{9401} \\ \frac{3451}{9401} & \frac{2501}{9401} & \frac{3449}{9401} \end{bmatrix}.$$

Finally, we must find the stationary probability vector of each subclass. Since

$$\mathbb{C}_1\mathbb{C}_2 = \begin{bmatrix} 0.86 & 0.14 \\ 0.80 & 0.20 \end{bmatrix},$$

we need a solution to

$$\begin{bmatrix} 0.14 & -0.80 \\ -0.14 & 0.80 \end{bmatrix} s_1 = \mathbf{0},$$

implying

$$s_1 = \begin{bmatrix} \frac{40}{47} \\ \frac{7}{47} \end{bmatrix}.$$

Based on this,

$$s_2^T = s_1^T \mathbb{C}_1 = \begin{bmatrix} \frac{23}{94} & \frac{71}{94} \end{bmatrix}.$$

The next four columns of $\mathbb{P}^{1000}$ are therefore

$$\begin{bmatrix} 0 & 0 & 0 & 0 \\ 0 & 0 & 0 & 0 \\ \frac{40}{47} & \frac{7}{47} & 0 & 0 \\ \frac{40}{47} & \frac{7}{47} & 0 & 0 \\ 0 & 0 & \frac{23}{94} & \frac{71}{94} \\ 0 & 0 & \frac{23}{94} & \frac{71}{94} \\ \frac{3786}{9401} \times \frac{40}{47} & \frac{3786}{9401} \times \frac{7}{47} & \frac{2759}{9401} \times \frac{23}{94} & \frac{2759}{9401} \times \frac{71}{94} \\ \frac{2501}{9401} \times \frac{40}{47} & \frac{2501}{9401} \times \frac{7}{47} & \frac{3449}{9401} \times \frac{23}{94} & \frac{3449}{9401} \times \frac{71}{94} \end{bmatrix}.$$

The remaining (transient) columns are identically equal to zero.   □

3.2 REVERSIBILITY

Suppose we observe an FMC in time *reverse*. Does it still behave as an FMC, and if so, what is the corresponding TPM?

The answer to the first question is NO whenever there are any transient states (absorption is a nonreversible process). If all states are recurrent (in which case we may consider each class separately), the answer is YES, but only when the process has reached its stationary state (i.e., after many transitions – this can also be arranged from the very start by using the *stationary* probabilities as the *initial* distribution – if the class is periodic, we use the *fixed* vector instead).

As we know, the TPM consists of the following probabilities: $\Pr(X_{n+1} = j \mid X_n = i) \equiv p_{ij}$. For the time-reversed FMC, we have

$$
\begin{aligned}
\overset{\circ}{p}_{ij} &\equiv \Pr(X_n = j \mid X_{n+1} = i) \\
&= \frac{\Pr(X_n = j \cap X_{n+1} = i)}{\Pr(X_{n+1} = i)} \\
&= \frac{\Pr(X_{n+1} = i \cap X_n = j)}{\Pr(X_{n+1} = i)} \\
&= \frac{\Pr(X_{n+1} = i \mid X_n = j)\Pr(X_n = j)}{\Pr(X_{n+1} = i)} \\
&= \frac{p_{ji} \cdot s_j}{s_i}.
\end{aligned}
$$

This implies taking the original TPM's transpose, then *multiplying* the first column of the resulting matrix by s_1, the second column by s_2, etc., and finally *dividing* the first row by s_1, the second row by s_2, etc.

Proposition 3.3. $\overset{\circ}{\mathbb{P}}$ *has the same stationary (fixed) distribution as the original* $\mathbb{P}$.

Proof.

$$
\sum_i s_i \overset{\circ}{p}_{ij} = \sum_i s_i \frac{p_{ji}}{s_i} s_j = \sum_i p_{ji} s_j = s_j.
$$

$\square$

Example 3.9. Based on $\mathbb{P} = \begin{bmatrix} 0.6 & 0.3 & 0.1 \\ 0.4 & 0.5 & 0.1 \\ 0.3 & 0.4 & 0.3 \end{bmatrix}$, we can construct $\overset{\circ}{\mathbb{P}}$ by

$$
\begin{bmatrix}
0.6 & 0.4 & 0.3 \\
0.3 & 0.5 & 0.4 \\
0.1 & 0.1 & 0.1
\end{bmatrix}
\begin{matrix}
\div 31 \\
\div 35 \\
\div 8
\end{matrix}
=
\begin{bmatrix}
0.6 & 0.32258\ldots & 0.07742\ldots \\
0.372 & 0.5 & 0.128 \\
0.3875 & 0.3125 & 0.3
\end{bmatrix}.
$$

$$
\begin{matrix}
\times & \times & \times \\
31 & 25 & 8
\end{matrix}
$$

$\diamondsuit$

A (single-class) Markov chain with $\overset{\circ}{\mathbb{P}} \equiv \mathbb{P}$ is called TIME-REVERSIBLE. A periodic FMC cannot be reversible when $\lambda > 2$; it can be reversible only when $\lambda = 2$, the two subclasses are of the same size, and the preceding condition is met. But typically, reversible FMCs are regular.

The previous example involved an FMC that was not reversible. On the other hand, the maze example *does* meet the reversibility condition since

$$
\overset{\circ}{\mathbb{P}} =
\begin{bmatrix}
\cdot & \cdot & \cdot & \frac{1}{2} & \cdot & \cdot \\
\cdot & \cdot & 1 & \cdot & \frac{1}{3} & \cdot \\
\cdot & \cdot & \cdot & \cdot & \cdot & \cdot \\
1 & \cdot & 1 & \cdot & \frac{1}{3} & \cdot \\
\cdot & \frac{1}{2} & \cdot & \cdot & \cdot & 1 \\
\cdot & \cdot & \cdot & \cdot & \frac{1}{3} & \cdot
\end{bmatrix}
\begin{matrix}
\div 1 \\
\div 2 \\
\div 1 \\
\div 2 \\
\div 3 \\
\div 1
\end{matrix}
$$

$$
\begin{matrix}
\times & \times & \times & \times & \times & \times \\
1 & 2 & 1 & 2 & 3 & 1
\end{matrix}
$$

is equal to the original $\mathbb{P}$.

We can easily construct a time-reversible TPM from any $\mathbb{P}$ by taking $\dfrac{\mathbb{P} + \overset{\circ}{\mathbb{P}}}{2}$ (another solution would be $\mathbb{P}\overset{\circ}{\mathbb{P}}$ or $\overset{\circ}{\mathbb{P}}\mathbb{P}$).

3.3 GAMBLER'S RUIN PROBLEM

We now return to our betting problem (Example 2.2), with the following modifications:

1. We are now playing against a casino (with unlimited resources), starting with i dollars and playing until broke (State 0) or until we have totaled N dollars (net win of $N - i$).
2. The probability of winning a single round is no longer exactly $\frac{1}{2}$, but (usually) slightly less, say p.

This time we abandon matrix algebra in favor of a different technique that utilizes the so-called DIFFERENCE EQUATIONS. This is possible due to a rather special feature of this (gambler's ruin) problem, where each state can directly transit into only one of its two *adjacent* states.

The trick is to keep N and p fixed, but consider all possible values of i (i.e., 0 to N) because they may all happen during the course of the game. We denote the probability of winning (reaching State N before State 0), given that currently we are in State i, by w_i.

Proposition 3.4. *The w_i must satisfy the following difference equations (Appendix 3.A):*

$$w_i = pw_{i+1} + qw_{i-1}$$

for $i \in \{1, 2, \ldots, N-1\}$, where $q \equiv 1 - p$.

Proof. We partition the sample space according to the outcome of the next round and use the formula of total probability. The probability of winning (ultimately) the game given that we are (now) in State i must equal the probability of winning the next round, multiplied by the probability of winning the game when starting with $i + 1$ dollars, plus the probability of losing the next round, multiplied by the probability of winning the game when starting with $i - 1$ dollars.   □

Furthermore (and quite obviously), $w_0 = 0$ and $w_N = 1$. We then must solve the $N - 1$ ordinary, linear equations for the $N - 1$ unknowns $w_1, w_2, \ldots, w_{N-1}$. This would normally be quite difficult due to the size of the problem, until we notice each equation involves only three consecutive w_i values, and is linear, with constant coefficients. There is a simple technique that can be used to solve such a set of equations in general (Appendix 3.A). The gist of it is as follows:

1. We substitute the following trial solution into (3.4):

$$w_i = \lambda^i,$$

getting $\lambda^i = p\lambda^{i+1} + q\lambda^{i-1}$ or

$$\lambda = p\lambda^2 + q$$

(a quadratic, so-called CHARACTERISTIC equation for λ).
2. We solve the characteristic equation; thus,

$$\lambda_{1,2} = \frac{1}{2p} \pm \sqrt{\frac{1}{4p^2} - \frac{q}{p}} = \frac{1}{2p} \pm \left| \frac{1 - 2p}{2p} \right| = \frac{q}{p} \text{ or } 1.$$

3. The general solution is then a linear combination of the two trial solutions with (at this point) arbitrary coefficients A and B:

$$w_i = A\left(\frac{q}{p}\right)^i + B.$$

4. Finally, we know $w_0 = 0$ and $w_N = 1$, or

$$A + B = 0$$

$$A\left(\frac{q}{p}\right)^N + B = 1,$$

implying

$$A = -B = \frac{1}{\left(\frac{q}{p}\right)^N - 1},$$

further implying

$$w_i = \frac{\left(\frac{q}{p}\right)^i - 1}{\left(\frac{q}{p}\right)^N - 1} \tag{3.4}$$

for any i.

We would now like to check our solution agrees with the results obtained in Example 2.2. Unfortunately, in that case, $p = \frac{1}{2}$, and our new formula yields $\frac{0}{0}$. But, with the help of L'Hopital's rule (calling $\frac{q}{p} \equiv x$),

$$\lim_{x \to 1} \frac{x^i - 1}{x^N - 1} = \lim_{x \to 1} \frac{ix^{i-1}}{Nx^{N-1}} = \frac{i}{N},$$

which is the correct answer.

Note when $\frac{q}{p} = \frac{19}{18}$ (a roulette with an extra zero), using $i = \frac{N}{2}$ (which would yield a fair game with $p = \frac{1}{2}$), we get, for the probability of winning, 36.80% when $N = 20$, only 6.28% when $N = 100$, and 2×10^{-12} (practically impossible) when $N = 1000$.

GAME'S EXPECTED DURATION

Let τ_i be the expected number of rounds the game will take (regardless of whether we win or lose). Then

$$\tau_i = p\tau_{i+1} + q\tau_{i-1} + 1$$

(the logic is similar to deriving the previous equation for w_i, except now we have to add 1 for the one round that has already been played). The boundary values of τ are $\tau_0 = \tau_N = 0$ (once we enter State 0 or State N, the game is over).

The solution is now slightly more complicated because $+1$ represents the nonhomogeneous term. We must first solve the homogeneous version of the equations (simply dropping $+1$), getting

$$\tau_i^{\text{hom}} = A \left(\frac{q}{p} \right)^i + B$$

as before. To find a solution to the full equation, we must add (to this homogeneous solution) a PARTICULAR solution to the complete equation.

As explained in Appendix 3.A, when the nonhomogeneous term is a constant, so is the particular solution, unless (which is our case) $\lambda = 1$ is one of the roots of the characteristic polynomial. Then we must use instead

$$\tau_i^{\text{part}} = c \cdot i,$$

where c is yet to be found, by substituting τ_i^{part} into the full equation; thus,

$$c \cdot i = p \cdot c \cdot (i + 1) + q \cdot c \cdot (i - 1) + 1,$$

which implies $c = \frac{1}{q-p}$. The full solution is then

$$\tau_i = A \left(\frac{q}{p} \right)^i + B + \frac{i}{q - p}.$$

To meet the boundary conditions, we solve

$$A + B = 0,$$

$$A \left(\frac{q}{p} \right)^N + B = \frac{-N}{q - p},$$

implying

$$A = -B = \frac{-N}{q - p} \cdot \frac{1}{\left(\frac{q}{p} \right)^N - 1}.$$

The final answer is

$$\tau_i = \frac{i - N \cdot w_i}{q - p}, \tag{3.5}$$

where w_i is given in (3.4).

This again will be indeterminate in the $p = \frac{1}{2}$ case, where, by L'Hopital's rule,

$$\lim_{x \to 1} \frac{i - N\frac{x^i - 1}{x^N - 1}}{\frac{1}{2}(x - 1)} = \lim_{x \to 1} \frac{i(x^N - 1) - N(x^i - 1)}{\frac{1}{2}(x^{N+1} - x^N - x + 1)}$$

$$= 2\frac{i\,N(N - 1) - N\,i(i - 1)}{(N + 1)N - N(N - 1)}$$

$$= i\,(N - i)$$

(note the second derivative in x had to be taken).

CORRESPONDING VARIANCE

Let Y_i be the number of rounds of the game to be played from now till its completion, given we are in State i. Then, by the same argument as before,

$$\mathbb{E}\left((Y_i - 1)^2\right) = p\,\mathbb{E}\left(Y_{i+1}^2\right) + q\,\mathbb{E}\left(Y_{i-1}^2\right),$$

where $Y_i - 1$, Y_{i+1}, and Y_{i-1} represent the game's duration *after* one round has been been played, in particular:

1. $Y_i - 1$; *before* the outcome is known,
2. Y_{i+1}; the outcome is a win, and
3. Y_{i-1}; the outcome is a loss.

$$\eta_i - 2\tau_i + 1 = p\,\eta_{i+1} + q\eta_{i-1},$$

where τ_i is given in (3.5) and $\eta_0 = \eta_N = 0$.

It is a bit more difficult to solve this equation in general since its nonhomogeneous part (namely, $1 - 2\tau_i$) is a sum of a constant, a term proportional to i, and another one proportional to $\left(\frac{q}{p}\right)^i$. Using the SUPERPOSITION PRINCIPLE, the particular solution can be built as a sum (superposition) of a general linear polynomial $c_0 + c_1 \cdot i$, further multiplied by i (since one root of the characteristic polynomial is 1), and $c_2\left(\frac{q}{p}\right)^i$, also multiplied by i, for the same reason. The details of this get rather tricky and tedious, so we will only work out the $p = \frac{1}{2}$ case and verify the general solution by MAPLE.

The $p = \frac{1}{2}$ assumption simplifies the equation since $\tau_i = i\,(N - i)$, the nonhomogeneous term is quadratic in i, and therefore

$$\eta_i^{\text{part}} = c_0 i^2 + c_1 i^3 + c_2 i^4 \tag{3.6}$$

(a general quadratic, multiplied by i^2, since now the characteristic polynomial has a *double* root of 1).

The equation itself now reads

$$\eta_i - 2i(N-i) + 1 = \frac{1}{2}\eta_{i+1} + \frac{1}{2}\eta_{i-1}.$$

Substituting (3.6), we get

$$2i^2 - 2Ni + 1 = c_0 + 3i\,c_1 + (6i^2 + 1)c_2.$$

This implies $c_2 = \frac{1}{3}$, $c_1 = -\frac{2}{3}N$, and $c_0 = \frac{2}{3}$.

The complete solution is thus

$$\eta_i = A + Bi + \frac{i^2}{3}(2 - 2iN + i^2),$$

where $A = 0$ and $B = \frac{N}{3}(N^2 - 2)$ to meet the two boundary conditions. Subtracting τ_i^2 we get

$$\begin{aligned}
\mathrm{Var}(Y_i) &= \frac{N}{3}(N^2 - 2)i + \frac{i^2}{3}(2 - 2iN + i^2) - i^2(N-i)^2 \\
&= \frac{1}{3}i(N-i)\left(i^2 + (N-i)^2 - 2\right)
\end{aligned}$$

symmetric under the $i \leftrightarrow N - i$ interchange (as expected).

One can derive that, in the general case,

$$\mathrm{Var}(Y_i) = \frac{N}{q-p}\left(\frac{4\left(\frac{q}{p}\right)^i \tau_i}{\left(\frac{q}{p}\right)^N - 1} - \frac{3Nw_i(1-w_i)}{q-p}\right) + \frac{4pq\tau_i}{(q-p)^2}.$$

We can verify this answer by

```
> w := n →  ((1-p)/p)^n - 1
            ─────────────── :
            ((1-p)/p)^N - 1
```

```
> simplify (w(n) − p · w(n + 1) − (1 − p) · w(n − 1));
```

$$0$$

```
> τ := n →  n − N · w(n)
           ───────────── :
             1 − 2 · p
```

```
> simplify (τ(n) − p · τ(n + 1) − (1 − p) · τ(n − 1) − 1);
```

$$0$$

```
> η := n →  N      ·  ⎛ 4 · ((1-p)/p)^n · τ(n)     3 · N · w(n) · (1 − w(n)) ⎞
           ─────        ─────────────────────  −  ───────────────────────────
           1 − 2 · p  ⎝   ((1-p)/p)^N − 1              1 − 2 · p              ⎠
```

$$+\frac{4 \cdot p \cdot (1-p) \cdot \tau(n)}{(1-2 \cdot p)^2} + \tau(n)^2 :$$

$$> simplify\,(\eta(n) - p \cdot \eta(n+1) - (1-p) \cdot \eta(n-1) + 1 - 2 \cdot \tau(n));$$

$$0$$

DISTRIBUTION OF THE GAME'S DURATION

Using a similar approach, one can derive the probability-generating function of the complete distribution of the number of rounds needed to finish the game.

Let $r_{i,n}$ be the probability that exactly n more rounds are needed, given that we are currently in State i. The corresponding difference equation reads

$$r_{i,n} = p \cdot r_{i+1,n-1} + q \cdot r_{i-1,n-1} \tag{3.7}$$

for $0 < n$ and $0 < i < N$. The trouble is now two indices are involved instead of the original one. To remove the second index, we introduce the corresponding probability-generating function (of the number of rounds to finish the game given that we are in State i); thus,

$$P_i(z) = \sum_{n=0}^{\infty} r_{i,n} z^n.$$

We now multiply (3.7) by z^n and sum over n, from 1 to ∞, to get

$$P_i(z) = pz\,P_{i+1}(z) + qz\,P_{i-1}(z),$$

which is a regular difference equation of the type we know how to solve. Quite routinely,

$$\lambda_{1,2} = \frac{1 \pm \sqrt{1-4pqz^2}}{2pz}, \tag{3.8}$$

and the general solution is

$$P_i(z) = A \cdot \lambda_1^i + B \cdot \lambda_2^i.$$

Imposing the conditions $P_0(z) = P_N(z) = 1$ (the number of remaining rounds is identically equal to zero in both cases), we get

$$A + B = 1$$

$$A \cdot \lambda_1^N + B \cdot \lambda_2^N = 1$$

or

$$A = \frac{1 - \lambda_2^N}{\lambda_1^N - \lambda_2^N},$$

$$B = \frac{1 - \lambda_1^N}{\lambda_2^N - \lambda_1^N}.$$

The final answer is

$$P_i(z) = \frac{\lambda_1^i(1 - \lambda_2^N) - \lambda_2^i(1 - \lambda_1^N)}{\lambda_1^N - \lambda_2^N}.$$

This is easily expanded by MAPLE, enabling us to extract the individual probabilities with λ_1 and λ_2 from (3.8):

> $\lambda_1 := \dfrac{1 + \sqrt{1 - 4 \cdot p \cdot (1 - p) \cdot z^2}}{2 \cdot p \cdot z}$:

> $\lambda_2 := \dfrac{1 - \sqrt{1 - 4 \cdot p \cdot (1 - p) \cdot z^2}}{2 \cdot p \cdot z}$:

> $P := \dfrac{\lambda_1^i \cdot (1 - \lambda_2^N) - \lambda_2^i \cdot (1 - \lambda_1^N)}{\lambda_1^N - \lambda_2^N}$:

> $(p, N, i) := \left(\dfrac{1}{2}, 20, 10\right)$: {This corresponds to a fair game.}

> $aux := series\,(P, z, 400)$:

> $pointplot\,([seq\,([2 \cdot i, coeff\,(aux, z, 2 \cdot i)], i = 5..190)])$;

3.A SOLVING DIFFERENCE EQUATIONS

We explained in Sect. 3.3 how to solve a *homogeneous* difference equation. We present the following examples to supplement that explanation.

Example 3.10. To solve

$$3a_{i+1} - 4a_i + a_{i-1} = 0,$$

we solve its characteristic polynomial $3\lambda^2 - 4\lambda + 1 = 0$, yielding $\lambda_{1,2} = 1, \frac{1}{3}$. This implies the general solution is

$$a_i = A + B\left(\frac{1}{3}\right)^i.$$

$\diamond$

Example 3.11. The equation

$$a_{i+1} + a_i - 6a_{i-1} = 0$$

results in $\lambda_{1,2} = 2, -3$, implying

$$a_i = A \cdot 2^i + B \cdot (-3)^i$$

for the general solution. If, furthermore, $a_0 = 8$ and $a_{10} = 60073$ (boundary conditions), we get, by solving

$$A + B = 8,$$
$$2^{10}A + (-3)^{10}B = 60073,$$

$A = 4$ and $B = 4$. The specific solution (solving both the equation and boundary conditions) is thus

$$a_i = 4\left(2^i + (-3)^i\right).$$

$\diamond$

When the two roots of a characteristic polynomial are identical (having a DOUBLE ROOT), we build the general solution in the following manner:

$$a_i = A\lambda^i + Bi\lambda^i.$$

Example 3.12. The equation

$$a_{i+1} - 4a_i + 4a_{i-1} = 0$$

results in $\lambda_{1,2} = 2, 2$, and the following general solution:

$$a_i = A \cdot 2^i + B \cdot i \cdot 2^i$$

(verify $i \cdot 2^i$ satisfies the equation).

$\diamond$

NONHOMOGENEOUS VERSION

When an equation has a NONHOMOGENEOUS part (usually placed on the right-hand side) that is a simple polynomial in i, the corresponding PARTICULAR solution can be built using a polynomial of the same degree with UNDETERMINED COEFFICIENTS. When, in addition to this, 1 is a single root of the characteristic polynomial, this trial solution must be further multiplied by i. The full solution is obtained by adding the particular solution to the general solution of the homogeneous version of the equation (obtained by dropping its nonhomogeneous term).

Example 3.13. (Particular solution only)

$$a_{i+1} + a_i - 6a_{i-1} = 3$$

requires $a_i^{\text{part}} = c$, implying $c = -\frac{3}{4}$.

Similarly, for

$$a_{i+1} + a_i - 6a_{i-1} = 2i + 3$$

we use $a_i^{\text{part}} = c_0 + c_1 \cdot i$, substitute it into the equation, and get

$$-4c_0 + c_1(7 - 4i) = 2i + 3,$$

which implies $c_1 = -\frac{1}{2}$ and $c_0 = -\frac{13}{8}$. The particular solution is thus

$$a_i^{\text{part}} = -\frac{1}{2} - \frac{13i}{8}.$$

$\diamond$

Example 3.14. (Single $\lambda = 1$ root)

$$3a_{i+1} - 4a_i + a_{i-1} = 5i - 2$$

requires $a_i^{\text{part}} = c_0 i + c_1 i^2$. Substituted into the previous equation, this yields

$$2c_0 + (4i + 4)c_1 = 5i - 2.$$

Thus, clearly, $c_1 = \frac{5}{4}$ and $c_0 = -\frac{7}{2}$.

$\diamond$

Finally, if the nonhomogeneous term has a form of $C \cdot \theta^i$, where θ is a constant distinct from all roots of the characteristic polynomial, then $a_i^{\text{part}} = c \cdot \theta^i$.

Example 3.15.

$$a_{i+1} + a_i - 6a_{i-1} = 5 \cdot \left(\frac{1}{2}\right)^i$$

Substituting $c \cdot \left(\frac{1}{2}\right)^i$ into this equation and canceling out $\left(\frac{1}{2}\right)^i$ yields $\frac{c}{2} + c - 6 \cdot 2c = 5$, implying $c = -\frac{10}{21}$. $\diamond$

When θ coincides with one of the roots, we must further multiply the trial solution by i.

Remark 3.1. This general approach works even when the two roots are complex (the A and B are then complex conjugates of each other whenever a real solution is desired).

COMPLEX-NUMBER ARITHMETIC

Briefly, for the two complex numbers $x = a + ib$ and $y = c + id$:
Addition and subtraction are performed componentwise:

$$(a + ib) + (c + id) = (a + c) + (b + d)i,$$

e.g.,

$$(3 - 5i) + (-2 + 3i) = 1 - 2i.$$

Multiplication uses the distributive law, and the property of the purely imaginary unit, namely, $i^2 = -1$

$$(a + bi) \times (c + di) = ac + (ad + bc)i + (ad)i^2 = (ac - ad) + (ad + bc)i.$$

For example,

$$(3 - 5i)(-2 + 3i) = -6 + 10i + 9i - 15i^2 = 9 + 19i.$$

Dividing two complex number utilizes the complex conjugate

$$\frac{a + bi}{c + di} = \frac{(a + bi)(c - di)}{(c + di)(c - di)} = \frac{(ac + ad) + (bc - ad)i}{c^2 + d^2},$$

e.g.,

$$\frac{3 - 5i}{-2 + 3i} = \frac{(3 - 5i)(-2 + 3i)}{(-2 + 3i)(-2 + 3i)} = \frac{-6 + 10i - 9i + 15i^2}{4 + 9} = -\frac{21}{13} + i\frac{1}{13}.$$

And, finally, raising a complex number to an integer power is best achieved by converting to polar coordinates. That is, since

$$a + bi = \sqrt{a^2 + b^2}\, e^{i \arctan(b,a)},$$

then

$$(a + bi)^n = \left(a^2 + b^2\right)^{\frac{n}{2}} e^{n \cdot i \arctan(b,a)},$$

where arctan uses the definition of MAPLE and $e^{i\theta} = \cos\theta + i\sin\theta$. For example (now using the usual $\tan^{-1}$ for hand calculation),

$$3 - 5i = \sqrt{3^2 + (-5)^2}\left(\cos\left(-\tan^{-1}\frac{3}{5}\right) + i\sin\left(-\tan^{-1}\frac{3}{5}\right)\right)$$

$$= -\sqrt{34} \cdot e^{-i\tan^{-1}\frac{3}{5}},$$

implying

$$(3 - 5i)^{27} = \sqrt{34}^{27} \cdot e^{-27i\tan^{-1}\frac{3}{5}}.$$

Example 3.16. To be able to use complex numbers (with their purely imaginary unit i) in this question, we replace a_i by a_n.

Consider

$$a_{n+1} + 2a_n + 10a_{n-1} = 0,$$

which corresponds to the characteristic equation $\lambda^2 + 2\lambda + 0 = 0$, resulting in $\lambda_{1,2} = -1 \pm \sqrt{1 - 10} = -1 \pm 3i$.

The *general solution* is therefore

$$a_n = A(-1 + 3i)^n + A^*(-1 - 3i)^n,$$

where A^* denotes the complex conjugate of A.

Now, adding the two initial conditions $a_0 = 3$ and $a_1 = -1$, we get

$$A + A^* = 3$$

and

$$A(-1 + 3i) + A^*(-1 - 3i) = -1.$$

The first equation implies the real part of A is $\frac{3}{2}$. Taking $A = \frac{3}{2} + ix$, the second equation yields $-3 - 6x = -1 \Rightarrow x = -\frac{1}{3}$. The complete solution is thus

$$a_n = \left(-\frac{i}{3}\right)(-1 + 3i)^n + \left(\frac{3}{2} + \frac{i}{3}\right)(-1 - 3i)^n.$$

This can also be written (in an explicitly real manner) as

$$a_n = \frac{\sqrt{85}}{3} \cdot 10^{\frac{n}{2}} \cdot \cos(\beta + n\theta),$$

where $\beta = \arctan\frac{2}{9}$ and $\theta = \arctan\frac{3}{\pi}$. $\diamondsuit$

EXERCISES

Exercise 3.1. Find $\overset{\circ}{\mathbb{P}}$ (the time-reversed TPM) of the following Markov chain:

$$\begin{bmatrix} 0 & 0 & 0.4 & 0.6 \\ 0 & 0 & 0.7 & 0.3 \\ 0.4 & 0.6 & 0 & 0 \\ 0.7 & 0.3 & 0 & 0 \end{bmatrix}.$$

Is the Markov chain reversible?

Exercise 3.2. Compute the expected number of transitions till absorption and the corresponding standard deviation, given that

$$\mathbb{P} = \begin{bmatrix} 1 & 0 & 0 & 0 & 0 \\ 0.3 & 0.2 & 0.5 & 0 & 0 \\ 0.2 & 0.4 & 0.4 & 0 & 0 \\ 0.2 & 0 & 0 & 0.6 & 0.2 \\ 0.1 & 0 & 0 & 0.5 & 0.4 \end{bmatrix}$$

and the process starts in State 5. Also, what is the expected number of visits to State 3?

Exercise 3.3. Is the Markov chain defined by

$$\begin{bmatrix} 0.13 & 0.15 & 0.20 & 0.11 & 0.18 & 0.23 \\ 0.09 & 0.19 & 0.17 & 0.14 & 0.30 & 0.11 \\ 0 & 0 & 1 & 0 & 0 & 0 \\ 0 & 0 & 0 & 1 & 0 & 0 \\ 0 & 0.33 & 0.31 & 0.12 & 0 & 0.24 \\ 0.33 & 0 & 0.12 & 0.31 & 0.24 & 0 \end{bmatrix}$$

lumpable as (a) 34|1256, (b) 123|456, (c) 12|3456, (d) 12|34|56, (e) 3|4|1256?

Exercise 3.4. For

$$
\mathbb{P} =
\begin{bmatrix}
0 & 1 & 0 & 0 & 0 & 0 & 0 \\
0.2 & 0.8 & 0 & 0 & 0 & 0 & 0 \\
0 & 0 & 0 & 0.3 & 0.7 & 0 & 0 \\
0 & 0 & 1 & 0 & 0 & 0 & 0 \\
0 & 0 & 1 & 0 & 0 & 0 & 0 \\
0 & 0 & 0 & 0.4 & 0 & 0 & 0.6 \\
0 & 0.5 & 0 & 0 & 0 & 0.5 & 0
\end{bmatrix}
$$

find (given the process starts in State 6):

(a) $\lim\limits_{n \to \infty} \mathbb{P}^{2n}$,
(b) $\lim\limits_{n \to \infty} \mathbb{P}^{2n+1}$,
(c) The expected number of transitions till absorption,
(d) The corresponding standard deviation.

Exercise 3.5. Consider a random walk through the following network of nodes (open circles are transient, solid circles are absorbing):

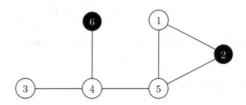

If the walk starts in Node 1, compute:

(a) The expected number of transitions till absorption and the corresponding standard deviation,
(b) The probability of being absorbed in Node 6.

Exercise 3.6. For

$$
\begin{bmatrix}
0.3 & 0.2 & 0.2 & 0.3 \\
0.3 & 0.4 & 0.3 & 0 \\
0.5 & 0.5 & 0 & 0 \\
1 & 0 & 0 & 0
\end{bmatrix}
$$

construct the PTMof the corresponding time-reversed Markov chain. Is this process reversible?

Exercise 3.7. Find the fixed vector of the following TPM:

$$
\begin{bmatrix}
0 & 0 & 0.2 & 0.4 & 0.4 \\
0 & 0 & 0.3 & 0.3 & 0.4 \\
0.2 & 0.8 & 0 & 0 & 0 \\
0.6 & 0.4 & 0 & 0 & 0 \\
0.3 & 0.7 & 0 & 0 & 0
\end{bmatrix}.
$$

If the process starts in State 1, what is the probability of reaching State 2 before State 3?

Exercise 3.8. Using the $\mathbb{P}$ of Exercise 2.14:

(a) Calculate the probability of never visiting State 1 given that the process starts in State 3.
(b) Determine the percentage of time the process will spend in State 5 if continued indefinitely.

Exercise 3.9. Consider

$$
\mathbb{P} =
\begin{bmatrix}
1 & 0 & 0 & 0 \\
0.2 & 0.5 & 0.2 & 0.1 \\
0 & 4 & 0.5 & 0.1 \\
0.1 & 0.2 & 0.4 & 0.3
\end{bmatrix}.
$$

If the initial state is chosen randomly (with the same probability for each of the four states), calculate the expected number of transitions till absorption and the corresponding standard deviation.

Exercise 3.10. Is the Markov chain defined by

$$
\begin{bmatrix}
0.21 & 0.27 & 0.07 & 0.14 & 0.31 \\
0.14 & 0.20 & 0.18 & 0.29 & 0.19 \\
0.23 & 0.18 & 0.40 & 0.07 & 0.12 \\
0.19 & 0.27 & 0.31 & 0.16 & 0.07 \\
0.11 & 0.18 & 0.20 & 0.19 & 0.32
\end{bmatrix}
$$

lumpable as (a) 14|3|25, (b) 14|2|35, (c) 134|25, (d) 14|235? Whenever it is, write down the new TPM.

CHAPTER 4
Branching Processes

BRANCHING PROCESSES are special Markov chains with infinitely many states. The states are nonnegative integers that usually represent the number of members of a population. Each of these members, before dying, leaves behind a random (possibly zero) number of offspring. This is repeated by the offspring themselves, from generation to generation, leading to either a population explosion or its ultimate extinction.

4.1 INTRODUCTION AND PREREQUISITES

Consider a population in which each individual produces, during his lifetime (which represents one time step and is called a GENERATION), a random number of offspring (according to a specific probability distribution). These in turn keep reproducing themselves in the same manner.

Examples:

1. Nuclear chain reaction (neutrons are the "offspring" of each atomic fission).
2. Survival of family names (carried by males) or of a new (mutated) gene.
3. In one-server queueing theory, customers arriving (and lining up) during the service time of a given customer can be, in this sense, considered that customer's "offspring" – this simplifies dealing with some tricky issues of queueing theory.

Intuitively, one can tell this model is not going to lead to a stable situation and that the ultimate fate of the process must be either total extinction or a population explosion. To verify that, let us first do some preliminaries.

J. Vrbik and P. Vrbik, *Informal Introduction to Stochastic Processes with Maple*, Universitext, DOI 10.1007/978-1-4614-4057-4_4,
© Springer Science+Business Media, LLC 2013

Compound Distribution

Suppose $X_1, X_2, \ldots, X_N$ is a random *independent* sample from a certain distribution, where N itself is a random variable (having its own distribution on nonnegative integers). For example, N may be the number of people stopping at a service station in a day, and X_i values are the amounts of gas they purchased.

For simplicity (sufficient for our purpose), we will assume the distribution of X_i is of a discrete (integer-valued) type and that its probability-generating function (PGF) is $P_X(s)$ (Appendix 4.A). Similarly, the PGF of the distribution of N is

$$P_N(s) = \sum_{k=0}^{\infty} \Pr(N = k) \cdot s^k.$$

We would now like to find the PGF of $S_N \equiv X_1 + X_2 + \cdots + X_N$ (the total purchases), say, $H(s)$. We know that

$$
\begin{aligned}
H(s) &= \sum_{k=0}^{\infty} \Pr(S_N = k) \cdot s^k \\
&= \sum_{k=0}^{\infty} \sum_{j=0}^{\infty} \Pr(S_N = k \mid N = j) \Pr(N = j) \cdot s^k \\
&= \sum_{j=0}^{\infty} \Pr(N = j) \sum_{k=0}^{\infty} \Pr(S_j = k) \cdot s^k \\
&= \sum_{j=0}^{\infty} \Pr(N = j) P_X^j(s) \\
&= P_N(P_X(s)).
\end{aligned}
$$

The PGF of S_N is thus a COMPOSITION of the PGF of N and that of X_i (note a composition of two functions is a noncommutative operation); the result is called a COMPOUND distribution.

One can easily show the corresponding mean is

$$\mathbb{E}(S_N) = P_N'(P_X(s)) \cdot P_X'(s)\big|_{s=1} = \mu_N \cdot \mu_X.$$

Similarly,

$\mathrm{Var}(S_N)$

$$
\begin{aligned}
&= P_N''(P_X(s)) \cdot P_X'(s)^2 + P_N'(P_X(s)) \cdot P_X''(s)\big|_{s=1} + \mathbb{E}(S_N) - \mathbb{E}(S_N)^2 \\
&= (\sigma_N^2 + \mu_N^2 - \mu_N) \cdot \mu_X^2 + (\sigma_X^2 + \mu_X^2 - \mu_X) \cdot \mu_N + \mu_N \cdot \mu_X - \mu_N^2 \cdot \mu_X^2 \\
&= \sigma_N^2 \cdot \mu_X^2 + \sigma_X^2 \cdot \mu_N.
\end{aligned}
$$

4.2 GENERATIONS OF OFFSPRING

We assume a branching process starts with a single individual (Generation 0), that is, $Z_0 \equiv 1$ (the corresponding PGF is thus equal to s). He (and ultimately all of his descendants) produces a random number of offspring, each according to a distribution whose PGF is $P(s)$. This is thus the PGF of the number of members of the first generation (denoted by Z_1).

The same Z_1 becomes the N for producing the next generation with Z_2 members since

$$Z_2 \equiv X_1 + X_2 + \cdots + X_{Z_1},$$

where the individual X_i are *independent* of each other. To get the PGF of Z_2, we must compound $P(s)$ (N's PGF) with the *same* $P(s)$ (the PGF of the X_i), getting $P(P(s))$.

Similarly, Z_2 is the effective N for creating the next generation, namely,

$$Z_3 \equiv X_1 + X_2 + \cdots + X_{Z_2},$$

whose PGF is thus the composition of $P(P(s))$ and another $P(s)$ (of the individual X), namely, $P(P(P(s))) \equiv P_{(3)}(s)$. Note the X adding up to Z_3 are different and *independent* of the X that have generated Z_2 – avoiding the need for an extra clumsy label, that is, $Z_3 \equiv X_1^{(3)} + X_2^{(3)} + \cdots$.

In general, the PGF of the number of members of the mth generation, i.e., of

$$Z_m \equiv X_1 + X_2 + \cdots + X_{Z_{m-1}},$$

is the m-fold composition of $P(s)$ with itself $[P_{(m)}(s)$, by our notation]. In some cases this can be simplified, and we get an explicit answer for the distribution of Z_m.

Example 4.1. Suppose the number of offspring (of each individual, in every generation) follows the modified (counting the failures only) geometric distribution with $p = \frac{1}{2}$ [this means $P(s) = \frac{\frac{1}{2}}{1 - \frac{s}{2}} = \frac{1}{2-s}$]. We can easily find that

$$P(P(s)) = \frac{1}{2 - \frac{1}{2-s}} = \frac{2-s}{3-2s},$$

$$P(P(P(s))) = \frac{2 - \frac{1}{2-s}}{3 - \frac{2}{2-s}} = \frac{3-2s}{4-3s},$$

$$P_{(4)}(s) = \frac{4-3s}{5-4s},$$

etc., and deduce the general formula is

$$P_{(m)}(s) = \frac{m - (m-1)s}{(m+1) - ms},$$

which can be proved by induction. Note the last function has a value of 1 at $s = 1$ (as it should), and it equals $\frac{m}{m+1}$ at $s = 0$ (this gives a probability of $Z_m = 0$, that is, extinction during the first m generations). As $\lim_{m \to \infty} \frac{m}{m+1} = 1$, extinction is certain in the long run. $\diamond$

Proposition 4.1. *When the geometric distribution is allowed to have any (permissible) value of the parameter p, the corresponding $P(s) = \frac{p}{1-qs}$, and*

$$P_{(m)}(s) = p \cdot \frac{p^m - q^m - (p^{m-1} - q^{m-1})qs}{p^{m+1} - q^{m+1} - (p^m - q^m)qs}.$$

Proof. This can be verified by induction, that is, by checking $P_{(1)}(s) \equiv \frac{p}{1-qs}$, and

$$P_{(m)}(s) = \frac{p}{1 - qP_{(m-1)}(s)}.$$

$\square$

Note $P_{(m)}(1) = 1$ for any m. Since the $m \to \infty$ limit of

$$P_{(m)}(0) = p \cdot \frac{p^m - q^m}{p^{m+1} - q^{m+1}}$$

is 1 when $p > q$, ultimate extinction is certain in that case. When $p < q$, the same limit is equal to $\frac{p}{q}$ (extinction can be avoided with a probability of $1 - \frac{p}{q}$). But now comes a surprise: the $m \to \infty$ limit of $P_{(m)}(s)$ is 1 for all values of s in the $p > q$ case but equals $\frac{p}{q}$ for all values of s (except for $s = 1$, where it equals 1) in the $p < q$ case.

What do you think this strange (discontinuous) form of $P_{(\infty)}(s)$ is trying to tell us?

GENERATION MEAN AND VARIANCE

Based on the recurrence formula for computing $P_{(m)}(s)$, namely,

$$P_{(m)}(s) \equiv P\left(P_{(m-1)}(s)\right),$$

we can easily derive the corresponding formula for the expected value of Z_m by a simple differentiation and the chain rule:

$$P'_{(m)}(s) \equiv P'\left(P_{(m-1)}(s)\right) \cdot P'_{(m-1)}(s).$$

Substituting $s = 1$ yields

$$\mu_m = \mu \cdot \mu_{m-1}$$

[since $P_{(m-1)}(1) = 1$], where $\mu = \mathbb{E}(X_i)$ (all X_i have identical distributions) and $\mu_m \equiv \mathbb{E}(Z_m)$. Since μ_1 is equal to μ, the last formula implies that

$$\mu_m = \mu^m.$$

Note when $\mu = 1$, $\mathbb{E}(Z_m) = 1$ for all m [yet $\Pr(Z_m = 0) \xrightarrow{m \to \infty} 1$ since extinction is certain; this is not a contradiction, try to reconcile it].

Similarly, one more differentiation

$$P''_{(m)}(s) = P''(P_{(m-1)}(s)) \cdot \left(P'_{(m-1)}(s) \right)^2 + P'(P_{(m-1)}(s)) \cdot P''_{(m-1)}(s)$$

yields (after the $s = 1$ substitution)

$$\gamma_m = \gamma \cdot \mu^{2(m-1)} + \mu \cdot \gamma_{m-1},$$

where γ_m is the second factorial moment of Z_m and $\gamma \equiv \gamma_1 \ (= \sigma^2 - \mu + \mu^2$, where σ^2 is the variance of the X_i distribution).

We thus have to solve the difference equation

$$\gamma_m - \mu \cdot \gamma_{m-1} = \gamma \cdot \mu^{2(m-1)}$$

for the γ_m sequence. The homogeneous solution is

$$\gamma_m = A \cdot \mu^m,$$

and a particular solution must have the form

$$\gamma_m^{\text{part}} = B \cdot \mu^{2m}.$$

Substituted into the original equation, this yields (after canceling μ^{2m})

$$B - B \frac{\mu}{\mu^2} = \frac{\gamma}{\mu^2} \Rightarrow B = \frac{\gamma}{\mu(\mu - 1)}.$$

The full solution is therefore

$$\gamma_m = A \cdot \mu^m + \gamma \cdot \frac{\mu^{2m-1}}{\mu - 1},$$

where A follows from $\gamma_1 \equiv \gamma$ (a single boundary condition), that is,

$$\gamma = A \cdot \mu + \gamma \cdot \frac{\mu}{\mu - 1} \Rightarrow A = -\frac{\gamma}{\mu(\mu - 1)}.$$

The final answer is

$$\gamma_m = \frac{\gamma}{\mu(\mu - 1)} \left(\mu^{2m} - \mu^m \right).$$

Converting this to the variance of Z_m we get

$$\mathrm{Var}(Z_m) = \frac{\gamma}{\mu(\mu-1)}\left(\mu^{2m}-\mu^m\right) - \mu^{2m}+\mu^m$$

$$= \left(\mu^{2m}-\mu^m\right)\left(\frac{\gamma}{\mu(\mu-1)}-1\right)$$

$$= \left(\mu^{2m}-\mu^m\right)\left(\frac{\sigma^2+\mu^2-\mu}{\mu(\mu-1)}-1\right)$$

$$= \sigma^2 \cdot \frac{\mu^{m-1}(\mu^m-1)}{\mu-1}.$$

When $\mu = 1$, we must use the limit of this expression (L'Hopital's rule), which results in $\sigma^2 (2m-1-(m-1)) = m \cdot \sigma^2$.

Example 4.2. Let us assume the offspring distribution is Poisson, with a mean $\lambda = 1$. What is the distribution of Z_{10} (number of members of Generation 10) and the corresponding mean and variance?

Solution.

```
> P := s → e^{s-1} :
> H := s :
> for i from 1 to 10 do
      H := P(H);
  end do:
> aux := series (H, s, 31) :
```

{There is a huge probability of extinction, namely:}
```
> coeff (aux, s, 0);
```
$$0.8418$$

{which we leave out of the following graph:}
```
> pointplot ([seq ([i, coeff (aux, s, i)], i = 1..30)]) ;
```

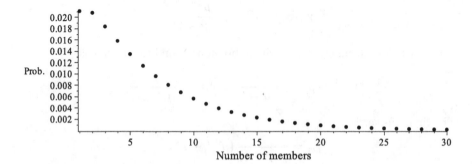

$$> \mu := \left. \frac{d}{ds} H \right|_{s=1} ;$$

$$\mu := 1.$$

$$> var := \left. \frac{d^2}{ds^2} H \right|_{s=1} + \mu - \mu^2;$$

$$var := 10.$$

{These are in agreement with our analytical formulas. Conditional mean and standard deviation (given the process is not yet extinct) may be more meaningful in this case:}

$$> \mu_c := \frac{\left. \frac{d}{ds} H \right|_{s=1}}{1 - H|_{s=0}};$$

$$\mu_c := 6.3197$$

$$> \sigma_c := \sqrt{ \frac{\left. \frac{d^2}{ds^2} H \right|_{s=1}}{1 - H|_{s=0}} + \mu_c - \mu_c^2 };$$

$$5.4385$$

◻

4.3 ULTIMATE EXTINCTION

We know $\lim_{m \to \infty} P_{(m)}(0)$ in general provides the probability of ultimate extinction of a branching process. This can be found by either computing the sequence $x_1 = P(0)$, $x_2 = P(x_1)$, $x_3 = P(x_2)$, and so on, until the numbers no longer change, or by solving

$$x_\infty = P(x_\infty). \tag{4.1}$$

In general, $x = 1$ is always a root of (4.1), but there might be another root in the $[0, 1)$ interval (if there is, it provides a value of x_∞; if not, $x_\infty = 1$ and ultimate extinction is certain).

Let us consider the geometric distribution with a parameter p whose PGF is $\frac{p}{1-qs}$. Equation (4.1) is equivalent to $qx^2 - x + p = 0$, with roots $\frac{1 \pm |p-q|}{2q}$

or 1 and $\frac{p}{q}$. When $p \geq \frac{1}{2}$, extinction is certain; for $p < \frac{1}{2}$, extinction happens with a probability of $\frac{p}{q}$ ($1 - \frac{p}{q}$ is thus the chance of indefinite survival). Note small p implies large "families" – the expected number of offspring is $\frac{q}{p}$.

When it is difficult to solve (4.1), the sequence x_1, x_2, x_3, ..., usually converges fast enough to reach a reasonable approximation to x_∞ in a handful of steps (with good knowledge of numerical analysis, one can speed up the convergence – but be careful not to end up with a wrong root!).

Example 4.3. Suppose the distribution for the number of offspring (of each member of a population) is Poisson, with $\lambda = 1.5$. Find the probability of ultimate extinction, assuming the population starts with a single member.

Solution. Since the corresponding $P(s) = e^{-1.5(1-s)}$, we get

$$x_1 = P(0) = e^{-1.5} = 0.2231$$
(the probability of being extinct, i.e., having no members,
in the first generation),

$$x_2 = P(1) = e^{-1.5(1-0.2231)} = 0.3118$$
(in the second generation),

$$x_3 = P(2) = e^{-1.5(1-0.3118)} = 0.3562$$
(in the third generation),

$$\vdots$$

$$x_{20} = P(19) = e^{-1.5(1-0.4172)} = 0.4172,$$

after which the value no longer increases (to this level of accuracy), being thus equal to the probability of ultimate extinction of the process. This can be done more easily with MAPLE.

```
> P := s → e^{1.5·(s−1)} :
> x_0 := 0 :
> for i from  0 to  20 do
      x_{i+1} := P (x_i);
  end do:
> convert (x, list);
```

$$[0, 0.2231, 0.3118, 0.3562, 0.3807, 0.3950, 0.4035, 0.4087, 0.4119, 0.4139,$$

$$0.4151, 0.4159, 0.4164, 0.4167, 0.4169, 0.4170, 0.4171, 0.4171, 0.4171,$$

$$0.4172, 0.4172, 0.4172]$$

{When only the ultimate value is needed, all we need is}

> $fsolve\,(x_\infty = P\,(x_\infty)\,,x_\infty = 0)\,;$

$$0.4172$$

{or graphically}
> $plot\,([P(s),s]\,,s = 0..1)\,;$

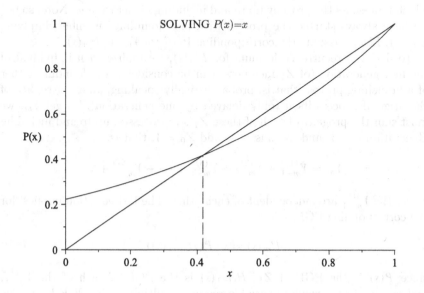

SOLVING $P(x)=x$

There is actually a very simple criterion for telling whether the process is headed for ultimate extinction or not: since $P(s)$ is CONVEX in the $[0, 1]$ interval [from $P''(s) \geq 0$, $P(0) = \Pr(X_i = 0) > 0$, and $P(1) = 1$], the graph of $y = P(s)$ can intersect the $y = s$ straight line (in the same interval) only when $P'(1) < 1$. But we know that $P'(1) = \mu$. The branching process thus becomes extinct with a probability of 1 whenever the average number of offspring (of a single individual) is less than *or equal to* 1.

TOTAL PROGENY

When ultimate extinction is certain, it is interesting to investigate the distribution of the *total* number of members of the branching process that will have ever lived (the so-called total PROGENY). As usual, if the distribution itself proves too complicated, we will settle for the corresponding mean and standard deviation.

Recall that in the context of queueing theory, total progeny represents the number of customers served during a BUSY PERIOD (which starts when a customer leaves the service and there is no one waiting – the last generation had no offspring).

We start by defining

$$Y_m = Z_0 + Z_1 + Z_2 + \cdots + Z_m,$$

which represents the progeny up to and including Generation m. Note, so far, we have always started the process with one "founding" member, implying $Z_0 \equiv 1$. Let us assume the corresponding PGF (of Y_m) is $H_m(s)$.

To derive a recurrence formula for $H_m(s)$, we realize each individual of the first generation (of Z_1 members) can be considered the founding father of a branching process that is, probabilistically speaking, an exact replica of the original process itself (only delayed by one generation). To get Y_m, we must sum the progeny of each of these Z_1 subprocesses, up to and including Generation $m - 1$, and we must also add $Z_0 = 1$, that is,

$$Y_m = Y_{m-1}^{(1)} + Y_{m-1}^{(2)} + Y_{m-1}^{(3)} + \cdots + Y_{m-1}^{(Z_1)} + 1,$$

where the $Y_{m-1}^{(i)}$ are independent of each other. The last equation implies, for the corresponding PGF,

$$H_m(s) = s \cdot P(H_{m-1}(s)) \tag{4.2}$$

since $P(s)$ is the PGF of Z_1, $H_{m-1}(s)$ is the PGF of each of the $Y_{m-1}^{(i)}$, and adding 1 to a random variable requires multiplying its PGF by s. The sequence starts with $H_0(s) = s$ (since the zeroth generation has only one member).

For $\mathbb{E}(Y_m) \equiv M_m$ we thus get, by differentiating (4.2) and setting $s = 1$,

$$M_m = 1 + \mu \cdot M_{m-1},$$

which results in $M_m = \dfrac{1 - \mu^{m+1}}{1 - \mu} \ (= 1 + \mu + \mu^2 + \cdots + \mu^m)$.

For the second factorial moment of Y_m (say S_m), we get, after another differentiation,

$$S_m = 2\mu \cdot M_{m-1} + (\sigma^2 - \mu + \mu^2) \cdot M_{m-1}^2 + \mu \cdot S_{m-1}.$$

Recalling $S_m = V_m + M_m^2 - M_m$ (where V_m is the corresponding variance), this yields

$$V_m + M_m^2 - M_m = 2\mu \cdot M_{m-1} + (\sigma^2 - \mu + \mu^2) \cdot M_{m-1}^2 + \mu \cdot (V_{m-1} + M_{m-1}^2 - M_{m-1})$$

or

$$V_m - V_{m-1} \cdot \mu = \mu \cdot M_{m-1} + (\sigma^2 + \mu^2) \cdot M_{m-1}^2 - M_m^2 + M_m = \sigma^2 \cdot M_{m-1}^2 \quad (4.3)$$

since

$$\mu \cdot M_{m-1} = M_m - 1$$

and

$$\mu^2 M_{m-1}^2 = M_m^2 - 2M_m + 1.$$

Solving the difference Eq. (4.3) for V_m yields

$$V_m = A \cdot \mu^m + \frac{\sigma^2}{(1-\mu)^2} \cdot \left(\frac{1 - \mu^{2m+1}}{1 - \mu} - 2m\mu^m \right).$$

Since $V_0 = 0$, we have

$$V_m = \sigma^2 \cdot \frac{1 - \mu^{2m+1}}{(1-\mu)^3} - \sigma^2 \cdot \frac{\mu^m(1 + 2m)}{(1-\mu)^2}. \quad (4.4)$$

The limit of this expression when $\mu \to 1$ is

$$\sigma^2 \frac{m}{3}(m + 1)(m + \tfrac{1}{2}).$$

Proposition 4.2. *The limit of the $H_m(s)$ sequence [which we call $H_\infty(s)$] represents the PGF of total progeny and must be a solution of*

$$H_\infty(s) = s \cdot P(H_\infty(s)). \quad (4.5)$$

Proof. Take the limit of (4.2) as $m \to \infty$. □

Example 4.4. In the case of a geometric distribution, $P(s) = \frac{p}{1-qs}$, and we need to solve

$$x = \frac{sp}{1 - qx},$$

which yields $x = \frac{1 \pm \sqrt{1 - 4pqs}}{2q}$. Since $H_\infty(0) = 0$ (total progeny cannot be 0), we must choose the minus sign for our solution; thus,

$$H_\infty(s) = \frac{1 - \sqrt{1 - 4pqs}}{2q}.$$

Note when $p > \frac{1}{2}$, $H_\infty(1) = 1$, but when $p < \frac{1}{2}$, $H_\infty(1) = \frac{p}{q}$ (why is the distribution "short" on probability?). ◇

More commonly, it is impossible to get a closed analytic solution for $H_\infty(s)$ [try solving $x = s \cdot e^{\lambda(x-1)}$, arising in the case of a Poisson distribution]; yet it is still possible to derive the values of $H'_\infty(1)$ and $H''_\infty(1)$, yielding the corresponding mean and variance. This is how it works:

Differentiating (4.5) results in

$$H'_\infty(s) = P(H_\infty(s)) + s \cdot P'(H_\infty(s)) \cdot H'_\infty(s),$$

implying (substitute $s = 1$) $\mathbb{E}(Y_\infty) = 1 + \mu \cdot \mathbb{E}(Y_\infty)$. Thus, we get the following very simple relationship between the expected value of the total progeny and the mean value of the number of offspring (of each individual):

$$\mathbb{E}(Y_\infty) = \frac{1}{1 - \mu}.$$

Differentiating (4.5) one more time yields

$$H''_\infty(s) = 2P'(H_\infty(s)) \cdot H'_\infty(s) + s \cdot P''(H_\infty(s)) \cdot \left(H'_\infty(s)\right)^2$$
$$+ s \cdot P'(H_\infty(s)) \cdot H''_\infty(s).$$

Substituting $s = 1$

$$H''_\infty(1) = \frac{2\mu}{1 - \mu} + \frac{\sigma^2 + \mu^2 - \mu}{(1 - \mu)^2} + \mu \cdot H''_\infty(1)$$

implies

$$H''_\infty(1) = \frac{\mu}{(1 - \mu)^2} + \frac{\sigma^2}{(1 - \mu)^3}.$$

The variance of Y_∞ is thus equal to

$$\mathrm{Var}(Y_\infty) = \frac{\mu}{(1 - \mu)^2} + \frac{\sigma^2}{(1 - \mu)^3} + \frac{1}{1 - \mu} - \frac{1}{(1 - \mu)^2}$$
$$= \frac{\sigma^2}{(1 - \mu)^3}.$$

(Note both of these formulas could have been derived – more easily – as a limit of the corresponding formulas for the mean and variance of Y_m.)

Remark 4.1. It is quite trivial to modify all our formulas to cover the case of Z_0 having any integer value, say N, instead of the usual 1 (effectively running N independent branching processes, with the same properties, in parallel). For a PGF and the probability of extinction, this means raising the corresponding $N = 1$ formulas to the power of N; for the mean and variance, the results get *multiplied* by N.

Example 4.5. Find the distribution (and its mean and standard deviation) of the progeny up to and including the tenth generation for a process with $Z_0 = 5$ and the offspring distribution being Poisson with $\lambda = 0.95$. Repeat for the *total* progeny (till extinction).

Solution.

```
> P := s → e^{0.95·(s−1)} :
> H := s :
> for i from  1 to  10 do
    H := s · P(H);
  end do:
> aux := series (H^5, s, 151) :
> pointplot ([seq ([i, coeff (aux, s, i)]) , i = 5..150]) ;
```

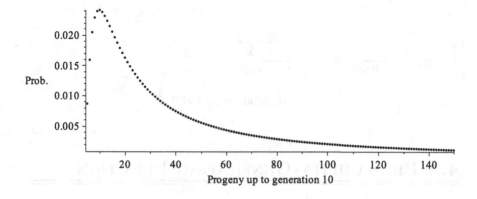

Progeny up to generation 10

$$> \mu := \left. \frac{\mathrm{d}}{\mathrm{d}s} H^5 \right|_{s=1} ;$$

$$\mu := 43.1200$$

$$> \sigma := \sqrt{ \left. \frac{\mathrm{d}^2}{\mathrm{d}s^2} \right|_{s=1} + \mu - \mu^2 } ;$$

$$\sigma := 34.1914$$

```
> for i from  1 to  200 do
    H := add (h_j · s^j, j = 1..i) ;
    aux := series (s · P(H), s, i + 1) ;
    h_i := solve (coeff (aux, s, i) = h_i, h_i) ;
  end do:
> aux := series (H^5, s, 201) :
> pointplot ([seq ([i, coeff (aux, s, i)]) , i = 5..200]) ;
```

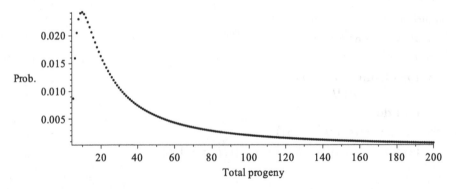

$$> \mu := \frac{5}{1 - 0.95}; \quad \sigma := \sqrt{\frac{5 \cdot 0.95}{(1 - 0.95)^3}};$$

$$\mu := 100.0000, \quad \sigma := 194.9359$$

4.A PROBABILITY-GENERATING FUNCTION

We recall a PGF of an (integer-valued) random variable X is defined by

$$P_X(s) \equiv \sum_{j=0}^{\infty} \Pr(X = j) \cdot s^j.$$

Note the following points:

1. $P_X(1) = 1$ since it yields the sum of all probabilities of a distribution.
2. This definition can include finitely many negative values of X when necessary.

Example 4.6. In the case of a modified (counting failures only) geometric distribution (Sect. 12.2), we get

$$P(s) = p + pqs + pq^2 s^2 + pq^3 s^3 + \cdots = \frac{p}{1 - qs}.$$

Clearly, $P'(s) = \sum_{j=0}^{\infty} \Pr(X = j) \cdot j \cdot s^{j-1}$, which implies

$$P'(1) = \mathbb{E}(X) = \mu_x$$

and, similarly,

$$P''(1) = \sum_{j=0}^{\infty} \Pr(X = j) \cdot j \cdot (j-1) = \mathbb{E}\left(X(X-1)\right)$$

(the second FACTORIAL MOMENT). The last two formulas further yield

$$\mathrm{Var}(X) = P''(1) + \mu_x - \mu_x^2.$$

Example 4.7. For a modified geometric distribution of the previous example, we get

$$\mu_x = \frac{pq}{(1-qs)^2}\bigg|_{s=1} = \frac{q}{p}$$

and

$$\mathrm{Var}(X) = \frac{2pq^2}{(1-qs)^3}\bigg|_{s=1} + \frac{q}{p} - \frac{q^2}{p^2} = \frac{qp + q^2}{p^2} = \frac{q}{p^2} = \frac{1}{p}\left(\frac{1}{p} - 1\right).$$

$\diamondsuit$

We also know, when X and Y are *independent*, the PGF of their sum is the *product* of the individual PGFs:

$$P_{X+Y}(s) = P_X(s) \cdot P_Y(s).$$

This can be easily generalized to a sum of three or more independent variables.

For convenience, we recall the PGF of a few common distributions:

Distribution	PGF
Poisson	$e^{-\lambda(1-s)}$
Binomial	$(q + ps)^n$
Negative binomial	$\left(\frac{ps}{1-qs}\right)^k$ or $\left(\frac{p}{1-qs}\right)^k$

depending on whether we are counting all trials or failures only, respectively (referring to the last box).

EXERCISES

Exercise 4.1. Consider a branching process with three initial members (Generation 0) and the following PGF for the number of offspring:

$$P(z) = \left(\frac{0.71 - 0.11z}{1 - 0.4z} \right)^2.$$

Compute:

(a) The probability that the last surviving generation (with at least one member) is Generation 4;
(b) The expected value and standard deviation of the number of members of Generation 5;
(c) The expected value and standard deviation of total progeny.

Exercise 4.2. Consider a branching process where the distribution of offspring is Poisson with a mean of 1.43. The process starts with four initial members (the 0th generation). Compute:

(a) The expected value of Y_5 (the process's progeny, up to and including Generation 5) and the corresponding standard deviation;
(b) The probability of the process's ultimate extinction;
(c) The probability that the process becomes extinct going from the second to the third generation (i.e., because the second generation has no offspring).

Exercise 4.3. Consider a branching process having the following distribution for the number of offspring:

X	0	1	2	3	4
Pr	0.31	0.34	0.20	0.10	0.05

and five initial members (in Generation 0). Compute:

(a) The probability of extinction within the first seven generations;
(b) The probability of ultimate extinction;
(c) The expected value and standard deviation of the number of members of Generation 7;
(d) The probability that Generation 7 has between 10 and 20 members (inclusive).

Exercise 4.4. Suppose each bacterium of a specific strain produces, during its lifetime, a random number of offspring whose distribution is binomial with $n = 5$ and $p = 0.15$. If we start with a culture containing 2,000 such bacteria, calculate the mean and the standard deviation of the total number of bacteria ever produced (including the original batch).

Exercise 4.5. Consider a branching process with four initial members (Generation 0) and the following PGF for the distribution of the number of offspring:

$$P(z) = \exp\left(\frac{9z - 9}{20 - 10z}\right).$$

Compute:

(a) The expected number of members this process will have had up to and including Generation 7 and the corresponding standard deviation;

(b) The expected number of generations till extinction and the corresponding standard deviation;

(c) The expected value of total progeny and the corresponding standard deviation.

Exercise 4.6. Consider a branching process where the number of offspring of each individual has a distribution with the following PGF:

$$\left(\frac{4}{5 - z}\right)^{4.5}.$$

Assuming the process starts with five individuals in Generation 0, compute:

(a) The probability of its ultimate extinction;

(b) The expected number of members of Generation 9 and the corresponding standard deviation;

(c) The probability Generation 9 will have: i) 20 members, or ii) 0 members.

Exercise 4.7. Suppose we change the PGF of the previous example to

$$\left(\frac{4}{5 - z}\right)^{3},$$

keeping the initial value of 5. Why must this process reach, sooner or later, extinction? Also compute:

(a) The expected time (measured in number of generations) till extinction and the corresponding standard deviation;

(b) The probability that extinction occurs at the third generation;

(c) The PGF of total progeny. Based on this, what is the probability that the total progeny exceeds 25?

Exercise 4.8. Consider a branching process where the number of offspring of each individual has the following distribution:

Number of offspring	0	1	2	3	4
Pr	0.40	0.35	0.15	0.07	0.03

Assuming the process starts with three individuals in Generation 0 compute:

(a) The probability of ultimate extinction;
(b) The probability that Generation 4 consists of more than five individuals;
(c) The expected value of the total progeny and the corresponding standard deviation;
(d) The probability that the total progeny exceeds 25.

CHAPTER 5
Renewal Theory

We now turn to processes that return repeatedly to their original state (e.g., betting \$1 on the flip of a coin and returning to the state where one has neither earned nor lost money, called "breaking even"). In this chapter we discuss only a special case of the renewal process: flipping a coin repeatedly until a specific pattern (e.g., HTHTH) is generated. Once achieved (the act of actual renewal) the game is reset and restarted. To make things more general, we will allow the probability of H to have any value: instead of flipping a coin, we can roll a die, for example, creating patterns of sixes and nonsixes. Eventually, we play two such patterns against each other.

5.1 PATTERN GENERATION

Suppose, in a sequence of Bernoulli trials, we try to generate a specific pattern of successes (S) and failures (F), for example, SFSFS. Let T be a random variable that counts the trials needed to succeed (for the first time).

If the sequence continues indefinitely, the pattern will be generated repeatedly. Let T_1, T_2, T_3, ...denote the number of trials needed to get the first, second, third, ...occurrence of the pattern. Assuming each new pattern must always be built *from scratch*, these random variables are *independent* and have the *same* distribution.

The reason we insist on always generating the same pattern from scratch is that we are actually interested in the number of trials it takes to generate the pattern *for the first time*. And this modification (generating the pattern from scratch *repeatedly*) happens to be the easiest way to deal with the issue.

Let f_n be the probability that the pattern is generated *for the first time* at the nth trial (the last letter of the pattern occurs on this trial). By definition,

J. Vrbik and P. Vrbik, *Informal Introduction to Stochastic Processes with Maple*, Universitext, DOI 10.1007/978-1-4614-4057-4_5,
© Springer Science+Business Media, LLC 2013

we take $f_0 = 0$. Note these probabilities define a probability distribution because they must add up to 1.

Let u_n be the probability that the pattern is generated at the nth trial, but not necessarily for the first time. By definition, $u_0 = 1$. Note the sum of these probabilities is infinite.

We will now find a relationship between the f_n and u_n probabilities. Let B be the event that our pattern occurs (is completed) at trial n (not necessarily for the first time) and $A_1, A_2, A_3, \ldots, A_n$ the event that the pattern is completed, *for the first time*, at trial 1, 2, 3, $\ldots$, n, respectively. If we add event A_0 (no occurrence of the pattern during the first n trials), then we have an obvious PARTITION of the sample space (consisting of all possible outcomes of the first n trials). The total-probability formula thus yields

$$\Pr(B) = \Pr(B \mid A_0) \Pr(A_0) + \Pr(B \mid A_1) \Pr(A_1) + \Pr(B \mid A_2) \Pr(A_2)$$
$$+ \cdots + \Pr(B \mid A_n) \Pr(A_n).$$

This can be rewritten as

$$u_n = u_n f_0 + u_{n-1} f_1 + u_{n-2} f_2 + \cdots + u_1 f_{n-1} + u_0 f_n \qquad (5.1)$$

[the first term on the right-hand side equals 0 in both formulas because $f_0 = 0$ and $\Pr(B \mid A_0) = 0$; the last term is also the same because $\Pr(B \mid A_n) = 1 = u_0$]. Note (5.1) is correct for any $n \geq 1$, but *not* (and this is quite important) for $n = 0$.

Multiplying each side of (5.1) by s^n and summing over n from 1 to ∞ we get $U(s) - 1$ on the left-hand side (since we are missing the $u_0 = 1$ term) and

$$u_0 f_0 + (u_1 f_0 + u_0 f_1)s + (u_2 f_0 + u_1 f_1 + u_0 f_2)s^2 + \cdots$$
$$= (u_0 + u_1 s + u_2 s^2 + \cdots)(f_0 + f_1 s + f_2 s^2 + \cdots)$$
$$= U(s) F(s)$$

on the right side where $U(s)$ is the generating function of the $u_0, u_1, u_2,$ $\ldots$ sequence (Appendix 5.A). Solving $U(s) - 1 = U(s) F(s)$ for $F(s)$ yields

$$F(s) = \frac{U(s) - 1}{U(s)}. \qquad (5.2)$$

As it happens, it is usually relatively easy to find the u_n probabilities and the corresponding $U(s)$. The previous formula thus provides a solution for $F(s)$.

RUNS OF r CONSECUTIVE SUCCESSES

Let the pattern we want to generate consist of r (a positive integer) successes (a RUN of length r). Let C be the event that all of the last r trials (out of n) resulted in a success (this does *not* imply our pattern was generated at Trial n, why?), and let B_{n-r+1}, B_{n-r+2}, $B_{n-r+3}, \ldots, B_n$ be the event that the pattern was generated (not necessarily for the first time) at Trial $n-r+1, n-r+2, n-r+3, \ldots, n$, respectively. Since the B_i events are *mutually exclusive* (right?) together with B_0 (the pattern was not completed during the last r trials), they constitute a *partition* of the sample space. We thus get

$$\begin{aligned} \Pr(C) = &\Pr(C \mid B_{n-r+1}) \Pr(B_{n-r+1}) + \Pr(C \mid B_{n-r+2}) \Pr(B_{n-r+2}) \\ &+ \Pr(C \mid B_{n-r+3}) \Pr(B_{n-r+3}) + \cdots + \Pr(C \mid B_n) \Pr(B_n) \\ &+ \Pr(C \mid B_0) \Pr(B_0). \end{aligned}$$

Now, since $\Pr(C) = p^r$, $\Pr(C \mid B_{n-r+i}) = p^{r-i}$, $\Pr(C \mid B_0) = 0$ and $\Pr(B_{n-r+i}) = u_{n-r+i}$, the last formula can be rewritten as

$$p^r = u_{n-r+1} \cdot p^{r-1} + u_{n-r+2} \cdot p^{r-2} + \cdots + u_{n-1} \cdot p + u_n \qquad (5.3)$$

(true only for $n \geq r$).

Multiplying each side by s^n and summing over n from r to ∞ results in

$$\begin{aligned} \frac{p^r s^r}{1-s} &= (U(s) - 1) s^{r-1} p^{r-1} + (U(s) - 1) s^{r-2} p^{r-2} + \cdots \\ &\quad + (U(s) - 1) sp + (U(s) - 1) \\ &= (U(s) - 1) \frac{1 - p^r s^r}{1 - ps}, \end{aligned}$$

which finally implies

$$U(s) - 1 = \frac{p^r s^r (1 - ps)}{(1-s)(1 - p^r s^r)}$$

or

$$U(s) = \frac{1 - s + q p^r s^{r+1}}{(1-s)(1 - p^r s^r)},$$

providing a solution for

$$F(s) = \frac{p^r s^r (1 - ps)}{1 - s + q p^r s^{r+1}}. \qquad (5.4)$$

Example 5.1. Having obtained the corresponding $F(s)$, we can find the probability of the first run of three consecutive heads being generated on the tenth flip of a coin as the coefficient of s^{10}, in an expansion of

$$\frac{s^3}{8} \left(1 - \frac{s}{2}\right) \left(1 - s + \frac{s^4}{16}\right)^{-1}$$

$$= \frac{s^3}{8} \left(1 - \frac{s}{2}\right) \left(1 + \left(s - \frac{s^4}{16}\right) + \left(s - \frac{s^4}{16}\right)^2 + \left(s - \frac{s^4}{16}\right)^3 + \cdots\right),$$

The terms that contribute are only

$$\frac{s^3}{8}\left(1-\frac{s}{2}\right)\left(\cdots+3s^2\left(-\frac{s^4}{16}\right)+4s^3\left(-\frac{s^4}{16}\right)+\cdots+s^6+s^7+\cdots\right),$$

yielding, for the final answer, $\frac{1}{8}\cdot(1-\frac{4}{16})-\frac{1}{16}\cdot(1-\frac{3}{16})=4.297\%$.

Similarly, the probability of needing more than 15 trials would be computed as the coefficient of s^{15} in an expansion of

$$\frac{1-F(s)}{1-s}=\left(1-\frac{s^3}{8}\right)\left(1-s+\frac{s^4}{16}\right)^{-1}.$$

The answer in this case is

$$1-\frac{12}{16}+\frac{\binom{9}{2}}{16^2}-\frac{\binom{6}{3}}{16^3}-\frac{1}{8}\left(1-\frac{1}{16^3}-\frac{9}{16}+\frac{\binom{6}{2}}{16^2}\right)=0.3238.$$

Of course, this can be done more easily using MAPLE.

```
> F := (s/2)^3 · (1.0 - s/2)/(1 - s + (s/2)^4);
> aux := series (F, s, 61) :
> coeff (aux, s, 10);
{Probability of needing exactly 10 trials:}

                    0.0430

> pointplot ([seq ([i, coeff(aux, s, i)], i = 3..60)]);
{Distribution of the number of trials needed to generate HHH:}
```

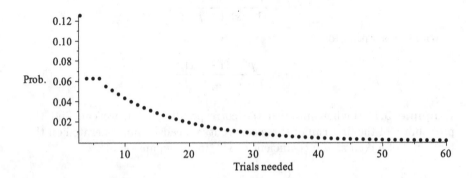

> *series* $\left(\dfrac{1-F}{1-s}, s, 16\right)$;

{Probability of needing more than i trials is the coefficients of s^i.}

$$1 + s + s^2 + 0.8750s^3 + 0.8125s^4 + 0.7500s^5 + 0.6875s^6 + 0.6328s^7$$

$$+0.58203s^8 + 0.5352s^9 + 0.4922s^{10} + 0.4526s^{11} + 0.4163s^{12}$$

$$+0.3828s^{13} + 0.3521s^{14} + 0.3238s^{15} + O(s^{16})$$

MEAN AND VARIANCE

By differentiating

$$F(s)(1 - s + qp^r s^{r+1}) = p^r s^r (1 - ps) \tag{5.5}$$

with respect to s and substituting $s = 1$, we obtain, for the corresponding expected value $\mu = F'(1)$,

$$\mu q p^r - 1 + (r + 1)qp^r = rqp^r - p^{r+1},$$

which implies

$$\mu = \frac{1 - p^r}{qp^r}.$$

Similarly, differentiating (5.5) *twice* (and substituting $s = 1$) yields

$$F''(1)qp^r + 2\mu[(r + 1)qp^r - 1] + (r + 1)rqp^r = r(r - 1)qp^r - 2rp^{r+1},$$

implying

$$F''(1)qp^r = 2\mu - 2(r + 1)(1 - p^r) - 2rp^r = 2\mu - 2(r + 1) + 2p^r.$$

The corresponding variance $\sigma^2 = F''(1) - \mu^2 + \mu$ is thus equal to

$$2\frac{1 - p^r}{(qp^r)^2} - 2\frac{r + 1}{qp^r} + \frac{2}{q} - \frac{1 - 2p^r + p^{2r}}{(qp^r)^2} + \frac{1 - p^r}{qp^r}$$

$$= \frac{1}{(qp^r)^2} - \frac{2r + 1}{qp^r} + \frac{1}{q} - \frac{1}{q^2}$$

$$= \frac{1}{(qp^r)^2} - \frac{2r + 1}{qp^r} - \frac{p}{q^2}.$$

Example 5.2. When we want to generate *three consecutive heads*, these formulas yield $\frac{1-\frac{1}{8}}{\frac{1}{16}} = 14$ for the expected number of trials and $\sqrt{16^2 - 7 \cdot 16 - 2}$ $= \sqrt{142} = 11.92$ for the corresponding standard deviation.

To get four consecutive heads:

$$\mu = \frac{1 - \frac{1}{16}}{\frac{1}{32}} = 30 \text{ and } \sigma = \sqrt{32^2 - 9 \cdot 32 - 2} = 27.09;$$

Two consecutive sixes (rolling a six-sided die):

$$\mu = \frac{1 - \frac{1}{36}}{\frac{5}{6} \cdot \frac{1}{36}} = 42 \text{ and } \sigma = \sqrt{\frac{216^2}{5^2} - 216 - \frac{6}{5^2}} = 40.62.$$

SECOND, THIRD, ETC. RUN OF r SUCCESSES

We have already found the PGF of the distribution for the number of trials to generate r consecutive successes for the first time (for the second, third, ..., time if we must always start from scratch).

If, on the other hand, each newly generated pattern is allowed to overlap with the previous one (making its generation easier), then we can generate the second (third, ...) occurrence of r consecutive successes by either

1. Achieving yet another success immediately (with the probability of p) or
2. Achieving a failure instead (thus breaking the run of consecutive successes and having to start from scratch, with a probability of q).

The distribution for the (extra) number of trials needed to generate the *second* (similarly the third, fourth, etc.) run of r consecutive successes (not necessarily from scratch) will thus be a mixture of two possibilities: the first one yields a value of 1 (with a probability of p), the second one results in 1 (for the extra failure) *plus* a random variable having a from-scratch distribution (with a probability of q). The overall PGF for the corresponding number of trials will thus be given by

$$ps + qsF(s).$$

Later on we discuss a general way of obtaining the PGF for the number of trials to generate any pattern.

Mean Number of Trials (Any Pattern)

Generation of a specific pattern can be completed in *any* number of trials (greater than or equal to its length). This means the corresponding recurrent event is APERIODIC (see the next section for an example of a periodic situation). In this (aperiodic) case, there is a simple way of finding the corresponding *mean* of the number of trials to generate the pattern (from scratch). We are also assuming the probability of generating the pattern (sooner or later) is 1.

One can show that, under these conditions, u_n must reach a fixed limit (say u_∞) when n becomes large (this is easy to understand intuitively). This u_∞ corresponds to the long-run proportion of trials in which the pattern is completed, implying

$$\mu = \frac{1}{u_\infty}.$$

The value of u_∞ can be established quite easily by going back to (5.3) and setting $n = \infty$ (which means each u_i of the equation becomes u_∞); thus,

$$p^r = u_\infty \cdot p^{r-1} + u_\infty \cdot p^{r-2} + \cdots + u_\infty \cdot p + u_\infty.$$

This implies

$$u_\infty = \frac{p^r}{1 + p + p^2 + \cdots + p^{r-1}} = \frac{p^r(1-p)}{1-p^r},$$

further implying

$$\mu = \frac{1 - p^r}{qp^r}$$

(which we already know to be the correct answer).

Example 5.3. Find the expected number of trials to generate SFSFS.

Solution. Assume the result of the last five trials (in a long sequence) was SFSFS. We know the probability of this happening is $pqpqp = p^3q^2$. This probability must equal (using the total-probability formula, according to where the last such from-scratch pattern was generated)

$$u_\infty + u_\infty \cdot qp + u_\infty \cdot q^2 p^2.$$

This corresponds to

Pr (the pattern *was* completed at the last of these five trials)

+ Pr (the pattern was completed on the third of the five trials)

+ Pr (the pattern was completed on the first of the five trials).

Note we get as many terms on the right-hand side of this equation as there are *matching overlaps* of the leading portion of this pattern with its trailing portion (when slid past itself, in one direction, including the full overlap). We thus obtain

$$p^3 q^2 = u_\infty (1 + pq + p^2 q^2),$$

implying

$$\mu = \frac{1 + pq + p^2 q^2}{p^3 q^2}.$$

When $p = \frac{1}{2}$, this results in 48 trials (on average).

Let us now find the corresponding variance. We must return to

$$p^3 q^2 = u_n + u_{n-2} \cdot pq + u_{n-4} \cdot p^2 q^2,$$

which implies (multiply the equation by s^n and sum over n from 5 to ∞ – note the equation is incorrect when $n \le 4$)

$$\frac{p^3 q^2 s^5}{1 - s} = (U(s) - 1) \cdot (1 + pqs^2 + p^2 q^2 s^4)$$

(because $u_1 = u_2 = u_3 = u_4 = 0$). Since

$$F(s) = \frac{U(s) - 1}{U(s)} \equiv \frac{1}{1 + \frac{1}{U(s)-1}},$$

we get

$$F(s) = \frac{1}{1 + (1 - s) \cdot Q(s)},$$

where

$$Q(s) = \frac{1 + pqs^2 + p^2 q^2 s^4}{p^3 q^2 s^5}.$$

Differentiating $F(s)$ yields

$$F'(s) = \frac{Q(s) - (1 - s) \cdot Q'(s)}{(1 + (1 - s) \cdot Q(s))^2},$$

implying the old result of $\mu = F'(1) = Q(1)$. One more differentiation (here, we also substitute $s = 1$) yields

$$F''(1) = 2Q'(1) + 2Q(1)^2.$$

The corresponding variance is thus equal to

$$2Q'(1) + \mu^2 + \mu = \mu^2 + \mu - 2\frac{5 + 3pq + p^2q^2}{p^3q^2}.$$

Using $p = \frac{1}{2}$, this has the value of $1980 \Rightarrow \sigma = 44.50$ (nearly as big as the mean).

Example 5.4. Find μ to generate the SSSFF pattern.

Solution. $p^3q^2 = u_\infty$ (no other overlap) implies

$$\mu = \frac{1}{p^3q^2}$$

($\mu = 32$, when $p = \frac{1}{2}$). Note it is easier to generate this pattern than to generate SFSFS since all of the occurrences of SSSFF count (there is no need to worry whether one was generated *from scratch* or not – there is no difference). On the other hand, some occurrences of SFSFS do not count as completing the SFSFS pattern *from scratch*. $\Box$

Using the new approach, it is now a lot easier to rederive the results for r consecutive successes. Since

$$Q(s) = \frac{1 - s^r p^r}{s^r p^r (1 - sp)},$$

we get immediately

$$\mu = \frac{1 - p^r}{p^r q},$$

and since

$$Q'(1) = \frac{-rp^r q + p^{r+1} - p^{2r+1}}{p^{2r} q^2},$$

the formula for the corresponding variance readily follows.

BREAKING EVEN

The same basic formula (5.2) also applies to the recurrent event of breaking even when the game involves betting one dollar repeatedly on a success in a Bernoulli sequence of trials. Note this is a PERIODIC situation (the period is 2 – one can break even only after an *even* number of trials).

The probability of breaking even after 2, 4, 6, ..., trials is equal to $\binom{2}{1}pq$, $\binom{4}{2}p^2q^2$, $\binom{6}{3}p^3q^3$, ..., respectively. The corresponding sequence-generating function (SGF) is thus

$$U(s) = 1 + \binom{2}{1}pqs^2 + \binom{4}{2}p^2q^2s^4 + \binom{6}{3}p^3q^3s^6 + \cdots,$$

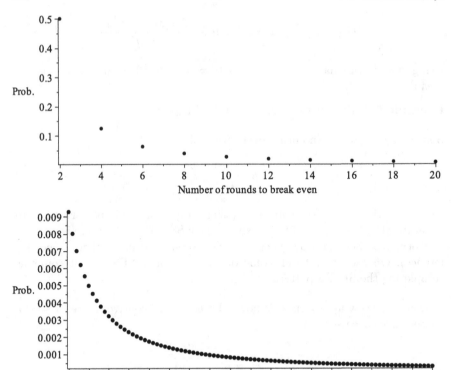

which is the expansion of $(1 - 4pqs^2)^{-\frac{1}{2}}$ (verify!). The PGF of the number of trials to reach the breakeven situation (for the first time, second time, etc. – here, we always start from scratch) is then

$$F(s) = 1 - \frac{1}{U(s)} = 1 - \sqrt{1 - 4pqs^2}.$$

Note $F(1) = 1 - |p - q|$, that is, it is equal to 1 only when $p = \frac{1}{2}$ (what happens in the $p \neq \frac{1}{2}$ case?).

The distribution of the number of rounds needed to break even (for the first time from now) can be established with MAPLE as follows:

```
> F := 1 - √(1. - s²) :
> aux := series (F, s, 201) :
> pointplot ([seq ([2 · i, coeff (aux, s, 2 · i)], i = 1..10)]) ;


> pointplot ([seq ([2 · i, coeff (aux, s, 2 · i)]) , i = 10..100]) ;
{continuation of the previous graph:}
```

Similarly, one can investigate the issue of two (or more) people playing this game (simultaneously, but independently of each other) reaching a point when they both (all) are winning (losing) the same amount of money. Interestingly, when $p = \frac{1}{2}$, this situation must recur with a probability of 1 only when the number of people is less than 4. But we will not go into details (our main topic remains pattern generation).

MEAN NUMBER OF OCCURRENCES

We now fix the number of Bernoulli trials at n and explore the number of occurrences of our recurrent event during this sequence (let us call the corresponding random variable N_n).

We know

$$\Pr(N_n \geq k) = \Pr(T_1 + T_2 + \cdots + T_k \leq n),$$

where $T_1, T_2, \ldots, T_k$ is the number of trials required to generate the first, second, $\ldots$, kth occurrence of a recurrent event, respectively. Since the PGF of $T_1 + T_2 + \cdots + T_k$ is $F(s)^k$, we know from (5.8) that the SGF (n being the sequence's index) of $\Pr(T_1 + T_2 + \cdots + T_k \leq n)$, and therefore of $\Pr(N_n \geq k)$ itself, is

$$\frac{F(s)^k}{1-s}.$$

This implies $\Pr(N_n = k) = \Pr(N_n \geq k) - \Pr(N_n \geq k+1)$ has, as its SGF (not PGF, since the index is n, not k)

$$\frac{1 - F(s)}{1-s} \cdot F(s)^k.$$

This can be converted to the SGF of the corresponding means $\mathbb{E}(N_n)$:

$$\frac{1-F(s)}{1-s} \sum_{k=0}^{\infty} k \cdot F(s)^k = \frac{1}{1-s} \cdot \frac{F(s)}{1-F(s)}$$

since $x + 2x^2 + 3x^3 + 4x^4 + \cdots = x \cdot (1 + x + x^2 + x^3 + \cdots)' = \frac{x}{(1-x)^2}$.

Example 5.5. When the recurrent event is defined as generating r consecutive successes, the previous formula results in

$$\frac{(1-ps)p^r s^r}{(1-s)^2(1-p^r s^r)}$$

due to (5.4) and also to

$$1 - F(s) = \frac{(1-s)(1-p^r s^r)}{1 - s + qp^r s^{r+1}}.$$

For $p = \frac{1}{2}$, $r = 3$, and $n = 10$, this yields, for the expected number of occurrences, a value of $\frac{627}{1024}$. This can be seen by rewriting

$$\frac{(1 - \frac{s}{2}) \cdot \frac{s^3}{8}}{(1 - s)^2 (1 - \frac{s^3}{8})}$$

in terms of partial fractions:

$$\frac{1}{14(1 - s)^2} - \frac{17}{98(1 - s)} + \frac{5 + 4s}{49(1 + \frac{s}{2} + \frac{s^2}{4})}.$$

In this form, it is easy to extract the coefficient of s^{10} of the first and second terms ($\frac{11}{14}$ and $-\frac{17}{98}$, respectively). The last term is equivalent (multiply the numerator and denominator by $s - \frac{1}{2}$) to

$$\frac{5 + \frac{3}{2}s - 2s^2}{49(1 - \frac{s^3}{8})} = \frac{5 + \frac{3}{2}s - 2s^2}{49} \cdot \left(1 + \frac{s^3}{8} + \frac{s^6}{64} + \frac{s^9}{512} + \cdots\right),$$

with the s^{10} coefficient equal to $\frac{3}{98 \times 512}$. The sum of the three coefficients yields $\frac{627}{1024}$. An alternate (but more general) approach is to express the last term as

$$\frac{5 + 4s}{49(1 + \frac{s}{2} + \frac{s^2}{4})} = \frac{\frac{5}{98} + \frac{11i}{98\sqrt{3}}}{1 + \frac{1+i\sqrt{3}}{4}s} + \frac{\frac{5}{98} - \frac{11i}{98\sqrt{3}}}{1 + \frac{1-i\sqrt{3}}{4}s} = \mathrm{Re}\left(\frac{\frac{5}{49} + \frac{11i}{49\sqrt{3}}}{1 + \frac{1+i\sqrt{3}}{4}s}\right)$$

and expand the denominator. These expansions are usually algebraically cumbersome and are best delegated to MAPLE (see below). $\diamondsuit$

Similarly, we can find the SGF of $\mathbb{E}(N_n^2)$ as being equal to

$$\frac{1 - F(s)}{1 - s} \sum_{k=0}^{\infty} k^2 \cdot F(s)^k = \frac{1}{1 - s} \cdot \frac{F(s)}{1 - F(s)} \cdot \frac{1 + F(s)}{1 - F(s)}$$

since

$$x + 2^2 x^2 + 3^2 x^2 + 4^2 x^2 + \cdots = x \cdot \left(x \cdot (1 + x + x^2 + x^3 + \cdots)'\right)' = \frac{x(1 + x)}{(1 - x)^3}.$$

This would enable us to compute the corresponding variance.

With the help of MAPLE, we can thus complete the previous example, as follows:

$$> F := \left(\frac{s}{2}\right)^3 \cdot \frac{1 - \frac{s}{2}}{1 - s + \left(\frac{s}{2}\right)^4} :$$

$$> series\left(\frac{F}{(1-F)\cdot(1-s)}, s, 11\right);$$

$$\frac{1}{8}s^3 + \frac{3}{16}s^4 + \frac{1}{4}s^5 + \frac{21}{64}s^6 + \frac{51}{128}s^7 + \frac{15}{32}s^8 + \frac{277}{512}s^9 + \frac{627}{1024}s^{10} + O\left(s^{11}\right)$$

$$> series\left(\frac{F}{(1-F)^2}\cdot\frac{1+F}{1-s}, s, 11\right);$$

$$\frac{1}{8}s^3 + \frac{3}{16}s^4 + \frac{1}{4}s^5 + \frac{23}{64}s^6 + \frac{59}{128}s^7 + \frac{73}{128}s^8 + \frac{357}{512}s^9 + \frac{851}{1024}s^{10} + O\left(s^{11}\right)$$

{The corresponding variance is thus:}

$$> \frac{851.}{1024} - \left(\frac{627}{1024}\right)^2;$$

$$0.4561$$

5.2 Two Competing Patterns

Suppose each of two players selects a specific pattern and then bets, in a series of Bernoulli trials, his/her pattern appears *before* the pattern chosen by the opponent. We want to know the individual probabilities of either player winning the game, the game's expected duration (in terms of the number of trials), and the corresponding standard deviation.

We first assume n trials of the random experiment have been completed and then define the following two sequences:

1. x_n is the probability that the first of the two patterns is completed at the nth trial, for the first time, without being preceded by the second pattern (and thus winning the game, at that point); we also take x_0 to be equal to 0.
2. y_n is, similarly, the reverse (the second pattern winning the game at the nth trial).

We also need the old sequence f_n (probability of the first pattern being completed, for the first time, at the nth trial, *ignoring* the second pattern, which may have been generated earlier) and its analog for the second pattern (generated for the first time at the nth trial, regardless of the first pattern) – this sequence will be called g_n.

Furthermore, we also need the following modification of f_n and g_n: let $\widehat{f}_n$ be the probability that the first pattern will need exactly n additional trials to be completed, for the first time, after the second trial is completed, *allowing the two patterns to overlap* (i.e., the first pattern may get some help from the second). We also take $\widehat{f}_0$ to have a value of 0.

Similarly (i.e., vice versa), we define $\widehat{g}_n$.

PROBABILITY OF WINNING

Let A be the event that the first pattern occurred, for the first time, at the nth trial (regardless of the second pattern), and let $B_1, B_2, B_3, \ldots, B_n$, B_0 define a partition of the sample space according to whether the second pattern won the game at the first, second, ..., nth trial or did not win the game at this point (B_0). Applying the total-probability formula yields

$$\Pr(A) = \Pr(B_1) \Pr(A \mid B_1) + \Pr(B_2) \Pr(A \mid B_2) + \cdots$$
$$+ \Pr(B_n) \Pr(A \mid B_n) + \Pr(A \cap B_0)$$

or

$$f_n = y_1 \cdot \widehat{f}_{n-1} + y_2 \cdot \widehat{f}_{n-2} + \cdots + y_n \cdot \widehat{f}_0 + x_n$$

correct for any integer $n \geq 0$.

Multiplying the previous equation by s^n and summing over n from 0 to ∞ results in

$$F(s) = Y(s) \cdot \widehat{F}(s) + X(s).$$

Clearly, the same argument can be made in reverse, obtaining

$$G(s) = X(s) \cdot \widehat{G}(s) + Y(s).$$

These two equations can be solved easily for

$$X(s) = \frac{F(s) - \widehat{F}(s)G(s)}{1 - \widehat{F}(s)\widehat{G}(s)} \tag{5.6}$$

and

$$Y(s) = \frac{G(s) - \widehat{G}(s)F(s)}{1 - \widehat{F}(s)\widehat{G}(s)}.$$

The probability that the first pattern wins the game is given by $x_1 + x_2 + x_3 + \cdots \equiv X(1)$. Unfortunately, substituting $s = 1$ into (5.6) results in $\frac{0}{0}$, and we need L'Hopital's rule to find the answer:

$$X(1) = \frac{\widehat{\mu} + v - \mu}{\widehat{\mu} + \widehat{v}},$$

where $\mu = F'(1)$ is the expected number of trials to generate Pattern 1 from scratch, $\widehat{\mu} = \widehat{F}'(1)$ is the expected number of trials to generate Pattern 1 starting from Pattern 2, and $v = G'(1)$ and $\widehat{v} = \widehat{G}'(1)$, with an analogous ($1 \leftrightarrow 2$) interpretation.

Similarly,

$$Y(1) = \frac{\widehat{v} + \mu - v}{\widehat{\mu} + \widehat{v}}.$$

Note $X(1) + Y(1) \equiv 1$ (as it should); also note when the two patterns are incompatible (no matching overlap such as, for example, a run of successes played against a run of failures), the formulas simplify [by removing the cap from $\widehat{F}(s)$, $\widehat{\mu}$, etc.].

Example 5.6. When betting r successes against ρ failures, the probability of r successes winning (appearing first) is

$$\frac{\frac{1-q^\rho}{pq^\rho}}{\frac{1-p^r}{qp^r} + \frac{1-q^\rho}{pq^\rho}} = \frac{(1-q^\rho)p^{r-1}}{p^{r-1} + q^{\rho-1} - p^{r-1}q^{\rho-1}}.$$

Or, to be more specific, betting two sixes against (a run of) 10 nonsixes gives us a

$$\frac{\left(1 - \left(\frac{5}{6}\right)^{10}\right) \cdot \frac{1}{6}}{\frac{1}{6} + \left(\frac{5}{6}\right)^9 - \frac{1}{6} \cdot \left(\frac{5}{6}\right)^9} = 42.58\%$$

chance of winning.   ◇

Example 5.7. When playing the SFFSS pattern against FFSSF, the situation is more complicated. Let μ_0, μ_1, μ_2, μ_3, and μ_4 be the expected number of trials needed (further needed) to build SFFSS from scratch, having S already, having SF, SFF, and, finally, SFFS, respectively (note $\mu = \mu_0$ and $\widehat{\mu} = \mu_2$). We can find these from the following set of equations:

$$\mu_0 = p \cdot \mu_1 + q \cdot \mu_0 + 1,$$
$$\mu_1 = p \cdot \mu_1 + q \cdot \mu_2 + 1,$$
$$\mu_2 = p \cdot \mu_1 + q \cdot \mu_3 + 1,$$
$$\mu_3 = p \cdot \mu_4 + q \cdot \mu_0 + 1,$$
$$\mu_4 = p \cdot 0 + q \cdot \mu_2 + 1,$$

where we already know the value of

$$\mu = \mu_0 = \frac{1 + p^2 q^2}{p^3 q^2}$$

(solving these equations would confirm this). The first equation implies $\mu_0 = \mu_1 + \frac{1}{p}$ and the second one $\mu_1 = \mu_2 + \frac{1}{q}$. We thus get

$$\widehat{\mu} = \mu_2 = \mu - \frac{1}{p} - \frac{1}{q}.$$

Similarly,

$$\nu_0 = p \cdot \nu_0 + q \cdot \nu_1 + 1,$$
$$\nu_1 = p \cdot \nu_0 + q \cdot \nu_2 + 1,$$
$$\nu_2 = p \cdot \nu_3 + q \cdot \nu_2 + 1,$$
$$\nu_3 = p \cdot \nu_4 + q \cdot \nu_1 + 1,$$
$$\nu_4 = p \cdot \nu_0 + q \cdot 0 + 1,$$

with

$$\nu = \nu_0 = \frac{1 + p^2 q^2}{p^2 q^3}$$

and

$$\widehat{\nu} = \nu_4 = 1 + \frac{1 + p^2 q^2}{pq^3}.$$

The probability of SFFSS winning over FFSSF is thus

$$\frac{\frac{1+p^2q^2}{p^2q^3} - \frac{1}{p} - \frac{1}{q}}{\frac{1+p^2q^2}{p^3q^2} - \frac{1}{p} - \frac{1}{q} + 1 + \frac{1+p^2q^2}{pq^3}} = \frac{\frac{1}{p^2q^3} - \frac{1}{p}}{\frac{1}{p^3q^2} + \frac{1}{pq^3}} = \frac{p - p^2q^3}{q + p^2}.$$

When $p = \frac{1}{2}$, this yields a $\frac{16-1}{16+8} = 62.5\%$ chance of winning for Player 1 (an enormous advantage). $\diamondsuit$

EXPECTED DURATION

From the previous section we know

$$H(s) = X(s) + Y(s) = \frac{F(s) + G(s) - \widehat{F}(s)G(s) - F(s)\widehat{G}(s)}{1 - \widehat{F}(s)\widehat{G}(s)}$$

is the PGF of the number of trials required to finish a game. We have already discussed how to obtain $F(s)$ and $G(s)$; $\widehat{F}(s)$ and $\widehat{G}(s)$ can be derived by a scheme similar to that for obtaining $\widehat{\mu}$ and $\widehat{\nu}$, for example,

$$F_0(s) = ps\, F_1(s) + qs\, F_0(s),$$
$$F_1(s) = ps\, F_1(s) + qs\, F_2(s),$$
$$F_2(s) = ps\, F_1(s) + qs\, F_3(s),$$
$$F_3(s) = ps\, F_4(s) + qs\, F_0(s),$$
$$F_4(s) = ps + qs\, F_2(s),$$

for the SFFSS pattern, where $F(s) = F_0(s)$ and $\widehat{F}(s) = F_2(s)$, and similarly for the FFSSF pattern.

We are usually not interested in the individual probabilities of the $H(s)$ distribution, only in the corresponding expected value obtained from $H'(1)$. To find this, we differentiate (twice, as it turns out to require)

$$H(s)\left(1 - \widehat{F}(s)\widehat{G}(s)\right) = F(s) + G(s) - \widehat{F}(s)G(s) - F(s)\widehat{G}(s), \qquad (5.7)$$

substituting $s = 1$ in the end. This yields

$$-2H'(1)\left(\widehat{\mu} + \widehat{v}\right) - \widehat{F}''(1) - 2\widehat{\mu}\widehat{v} - \widehat{G}''(1) = -\widehat{F}''(1) - 2\widehat{\mu}v - 2\mu\widehat{v} - \widehat{G}''(1)$$

(note the second derivatives cancel out), implying

$$H'(1) \equiv M = \frac{\widehat{\mu}v + \mu\widehat{v} - \widehat{\mu}\widehat{v}}{\widehat{\mu} + \widehat{v}} \equiv \frac{\frac{\mu}{\widehat{\mu}} + \frac{v}{\widehat{v}} - 1}{\frac{1}{\widehat{\mu}} + \frac{1}{\widehat{v}}}.$$

(In the incompatible case, the formula simplifies to $M = \frac{1}{\frac{1}{\mu} + \frac{1}{v}}$, which is half the harmonic mean of μ and v.)

Example 5.8. For the game where the SFFSS pattern is played against FFSSF, this yields

$$\frac{\frac{34}{30} + \frac{34}{18} - 1}{\frac{1}{30} + \frac{1}{18}} = 22.75 \text{ trials}$$

when $p = \frac{1}{2}$. $\diamondsuit$

To simplify our task, we will derive a formula for the variance V of the number of trials to complete the game only in the incompatible case [which implies $\widehat{F}(s) \equiv F(s)$, $\widehat{G}(s) \equiv G(s)$, $\widehat{\mu} \equiv \mu$, and $\widehat{v} \equiv v$]. This means (5.7) reduces to

$$H(s)\left(1 - F(s)G(s)\right) = F(s) + G(s) - 2F(s)G(s).$$

Differentiating three times yields

$$-3H''(1)\left(\mu + v\right) - 3M\left(F''(1) + 2\mu v + G''(1)\right)$$
$$- \left(F'''(1) + 3F''(1)v + 3G''(1)v + G'''(1)\right),$$

which reduces to

$$F'''(1) + G'''(1) - 2\left(F'''(1) + 3F''(1)v + 3G''(1)v + G'''(1)\right),$$

implying

$$H''(1) = \frac{v^2}{(\mu + v)^2} \cdot F''(1) + \frac{\mu^2}{(\mu + v)^2} \cdot G''(1) - 2M^2$$

(since $M = \frac{\mu \nu}{\mu + \nu}$). Replacing $H''(1)$ by $V - M + M^2$, $F''(1)$ by $\sigma_1^2 - \mu + \mu^2$, and $G''(1)$ by $\sigma_2^2 - \nu + \nu^2$ (where σ_1^2 and σ_2^2 are the individual variances of the number of trials to generate the first and second patterns, respectively), we get

$$V - M + M^2 = \frac{\nu^2}{(\mu + \nu)^2} \cdot (\sigma_1^2 - \mu + \mu^2) + \frac{\mu^2}{(\mu + \nu)^2} \cdot (\sigma_2^2 - \nu + \nu^2) - 2M^2,$$

implying

$$V = \frac{\nu^2}{(\mu + \nu)^2} \cdot \sigma_1^2 + \frac{\mu^2}{(\mu + \nu)^2} \cdot \sigma_2^2 - M^2 = P_1^2 \cdot \sigma_1^2 + P_2^2 \cdot \sigma_2^2 - M^2,$$

where P_1 (P_2) is the probability that the first (second) pattern wins the game.

Example 5.9. When playing 2 consecutive sixes against 10 consecutive non-sixes, the previous formula yields

$$\mu = \frac{1 - \left(\frac{1}{6}\right)^2}{\frac{5}{6} \cdot \left(\frac{1}{6}\right)^2} = 42,$$

$$\nu = \frac{1 - \left(\frac{5}{6}\right)^{10}}{\frac{1}{6} \cdot \left(\frac{5}{6}\right)^{10}} = 31.1504,$$

$$\sigma_1^2 = \frac{1}{\left(\frac{5}{6}\right)^2 \cdot \left(\frac{1}{6}\right)^4} - \frac{5}{\frac{5}{6} \cdot \left(\frac{1}{6}\right)^2} - \frac{\frac{1}{6}}{\left(\frac{5}{6}\right)^2} = 1650,$$

$$\sigma_2^2 = \frac{1}{\left(\frac{1}{6}\right)^2 \cdot \left(\frac{5}{6}\right)^{20}} - \frac{21}{\frac{1}{6} \cdot \left(\frac{5}{6}\right)^{10}} - \frac{\frac{5}{6}}{\left(\frac{1}{6}\right)^2} = 569.995.$$

The variance of the number of trials to complete this game thus equals

$$\left(\frac{31.1504}{42 + 31.1504}\right)^2 \cdot 1650 + \left(\frac{42}{42 + 31.1504}\right)^2 \cdot 569.995 - \left(\frac{42 \cdot 31.1504}{42 + 31.1504}\right)^2$$
$$= 167.231.$$

This translates into a standard deviation of 12.93 (the expected value of the game's duration is 17.89). $\diamond$

5.A SEQUENCE-GENERATING FUNCTION

Consider an infinite sequence of numbers, say $a_0, a_1, a_2, \ldots$. Its SGF is defined by

$$A(s) = \sum_{i=0}^{\infty} a_i s^i$$

(which is analogous to the PGF of a discrete probability distribution, except now the a_i do not need to be positive or add up to 1). For example, when all a_i are equal to 1, the corresponding SGF is $\frac{1}{1-s}$.

Example 5.10. What is the SGF of the following sequence: $a_0, a_0 + a_1,$ $a_0 + a_1 + a_2, \ldots$ (its ith term is defined by $c_i = \sum_{j=0}^{i} a_j$)?

Solution. Since $C(s) = a_0 + (a_0 + a_1)s + (a_0 + a_1 + a_2)s^2 + (a_0 + a_1 + a_2 + a_3)s^3 + \cdots$, we can see $C(s) - sC(s) = A(s)$, implying

$$C(s) = \frac{A(s)}{1-s}. \tag{5.8}$$

$\square$

When $A(s)$ happens to be a PGF of a random variable X, $C(s)$ would generate the following sequence: $\Pr(X \le 0)$, $\Pr(X \le 1)$, $\Pr(X \le 2)$, $\Pr(X \le 3)$, $\ldots$; these are the values of the corresponding *distribution function* $F(0)$, $F(1)$, $F(2)$, $\ldots$.

Proposition 5.1. *The sequence*

$$a_0 + b_0, \ a_1 + b_1, \ a_2 + b_2, \ a_3 + b_3, \ \ldots$$

has $A(s) + B(s)$ as its SGF. Thus, when $P(s)$ is a PGF, $\frac{1-P(s)}{1-s}$ yields a SGF of the following sequence:

$$\Pr(X > 0), \ \Pr(X > 1), \ \Pr(X > 2), \ \ldots.$$

Proof. Notice $\frac{1}{(1-s)}$ is a generating function of the sequence 1, 1, 1, $\ldots$. Moreover, $\Pr(X > k) = 1 - \Pr(X \le k)$. $\square$

EXERCISES

Exercise 5.1. Consider betting repeatedly \$1 on the flip of a coin.

(a) What is the probability that breaking even *for the third time* will happen during the first 50 rounds?

(b) What is the expected number of times one will break even during the first 50 rounds and the corresponding standard deviation?

Exercise 5.2. Find the expected number of rolls (and the corresponding standard deviation) to generate the following patterns:

(a) Five consecutive nonsixes,
(b) 6EE6E ("6" means six, "E" means anything else).
(c) What is the probability that the latter pattern will take more than 50 flips to generate?

Exercise 5.3. Consider flipping a coin to generate the pattern HTHTHT. What is the expected number of flips (and the corresponding standard deviation) to generate this pattern for the third time, assuming either

(a) The three occurrences must not overlap or
(b) One can utilize any number of symbols of the previous occurrence to generate the next one.

Exercise 5.4. Calculate the probability of getting three consecutive sixes before eight consecutive nonsixes. What is the expected duration and the corresponding standard deviation of such a game? What is the probability that completing two such games will take fewer than 200 rolls?

Exercise 5.5. If the pattern HTTH is played against THT, find its probability of winning. Also find the expected duration of the game (in terms of the number of flips) and the corresponding standard deviation.

CHAPTER 6
Poisson Process

We investigate the simplest example of a process run in real (continuous) time, with the state space consisting of nonnegative integers (usually a count of arrivals at a store, gas station, library, etc.). The process, whose value at time t we denote by $N(t)$, can make a transition from state n *only* to state $n + 1$; it does so by an instantaneous jump at random times (individual customer arrivals).

6.1 BASICS

Let $N(t)$ denote the number of cars that arrive at a gas station randomly but at a constant average rate, λ, during time t. The graphical representation $N(t)$ is a straight line, parallel to x, that once in a while (at the time of each arrival) makes a discrete jump of one unit up the y scale (as illustrated in Fig. 6.1).

To find the distribution of $N(t)$, we introduce the following notation:

$$P_n(t) = \Pr\left(N(t) = n \mid N(0) = 0\right), \tag{6.1}$$

where $n = 0, 1, 2, \ldots$.

The random variables $N(t + s) - N(t)$ for any t and s positive are called INCREMENTS of the process. They are assumed to be *independent* of the past and present, and of each other (as long as their time intervals do not overlap). Furthermore, the distribution of $N(t + s) - N(t)$ depends only on s but not t (the *homogeneity* condition). This implies

$$\Pr\left(N(t + s) - N(t) = n \mid N(t)\right) = P_n(s),$$

regardless of the value of t.

J. Vrbik and P. Vrbik, *Informal Introduction to Stochastic Processes with Maple*, Universitext, DOI 10.1007/978-1-4614-4057-4_6,
© Springer Science+Business Media, LLC 2013

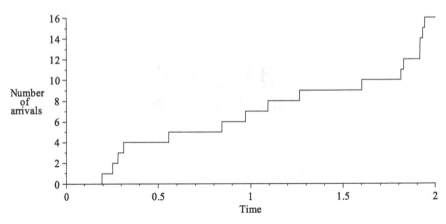

Fig. 6.1: Plot of $N(t)$; each step represents a new arrival

We now assume

$$\Pr\left(N(t+s) - N(t) = 1 \mid N(t) = i\right) = P_1(s) = \lambda \cdot s + o(s)$$

and

$$\Pr\left(N(t+s) - N(t) \geq 2 \mid N(t) = i\right) = \sum_{n=2}^{\infty} P_n(s) = o(s),$$

which imply

$$\Pr\left(N(t+s) - N(t) = 0 \mid N(t) = i\right) = P_0(s) = 1 - \lambda \cdot s + o(s),$$

where $o(s)$ is the usual notation for a function of s, say $f(s)$, such that $\lim_{s\to 0} \frac{f(s)}{s} = 0$. This normally means the Taylor expansion of $f(s)$ starts with the s^2 term (no absolute or linear term in s).

To find the $P_n(t)$ probabilities, we start with the following expansion, based on the formula of total probability:

$$P_n(t+s) = P_n(t) P_0(s) + P_{n-1}(t) P_1(s) + P_{n-2}(t) P_2(s) + \cdots + P_0(t) P_n(s).$$

From each side we then subtract $P_n(t)$, divide the result by s, and take the $s \to 0$ limit. This yields

$$\dot{P}_n(t) = -\lambda \cdot P_n(t) + \lambda \cdot P_{n-1}(t), \tag{6.2}$$

when $n \geq 1$, and

$$\dot{P}_0(t) = -\lambda \cdot P_0(t), \tag{6.3}$$

when $n = 0$, where the dot over P indicates differentiation with respect to t.

To solve this system of difference-differential equations, we introduce the following probability-generating function (PGF) of (6.1):

$$P(z,t) \equiv \sum_{n=0}^{\infty} P_n(t) \cdot z^n \tag{6.4}$$

(actually, a family of PGFs, one for each t). If we multiply (6.2) by z^n, sum over n from 1 to ∞, and add (6.3), we get

$$\dot{P}(z,t) = -\lambda \cdot P(z,t) + \lambda \cdot z \cdot P(z,t) = -\lambda(1-z)P(z,t).$$

Since the solution to

$$y' = a \cdot y$$

is

$$y(x) = c \cdot e^{a \cdot x},$$

we can solve the previous equation accordingly:

$$P(z,t) = c \cdot e^{-\lambda(1-z)t}.$$

We also know $P(z,0) = 1$ because $P_0(0) = 1$ and $P_n(0) = 0$ for $n \geq 1$ (the process starts in State 0). This means $c = 1$ and

$$P(z,t) = e^{-\lambda(1-z)t} = e^{-\lambda \cdot t} \cdot e^{\lambda z t}.$$

Expanded in z, this yields

$$e^{-\lambda \cdot t} \cdot \left(1 + \lambda t z + \frac{(\lambda t)^2}{2} z^2 + \frac{(\lambda t)^3}{3!} z^3 + \frac{(\lambda t)^4}{4!} z^4 + \cdots \right),$$

which further implies

$$P_n(t) = \frac{(\lambda t)^n}{n!} e^{-\lambda \cdot t}. \tag{6.5}$$

The distribution of $X(t)$ is thus Poisson, with a mean value of $\lambda \cdot t$.

To simulate the Poisson process, one can generate the interarrival times based on the following proposition.

Proposition 6.1. *The interarrival times of the Poisson process, denoted by V_i, are independent random variables from an exponential distribution with a mean of $\frac{1}{\lambda}$.*

Proof.

$$1 - F_{V_1}(t) \equiv \Pr(V_1 > t) = \Pr(N(t) = 0 \,|\, N(0) = 0) = e^{-\lambda \cdot t}$$

since the process is time homogeneous and Markovian, given the first arrival has just occurred, the time until the next arrival has, independently, the same distribution as V_1, etc. $\square$

There is yet another, more elegant, way of generating the Poisson process during a fixed time interval $(0, t)$.

Proposition 6.2. *To do that, first generate the* total number *of arrivals using a Poisson distribution with mean equal to λt, then draw the corresponding arrival times, uniformly and independently from the $[0, t]$ interval.*

Proof. This follows from the fact that the joint probability density function (PDF), $f(t_1, t_2, \ldots, t_n)$, of the arrival times $T_1, T_2, \ldots, T_n$, *given that* $N(t)=n$, equals

$$\lim_{h \to 0} \frac{\Pr\left(T_1 \in [t_1, t_1 + h) \cap T_2 \in [t_2, t_2 + h) \cap \cdots \cap T_n \in [t_n, t_n + h)\right)}{h^n \cdot \Pr\left(N(t) = n\right)}$$

$$= \lim_{h \to 0} \frac{(\lambda h + o(h))^n \, e^{-\lambda(t - nh)}}{h^n \cdot \dfrac{(\lambda t)^n}{n!} e^{-\lambda t}}$$

$$= \frac{n!}{t^n}$$

for $0 < t_1 < t_2 < \cdots < t_n < t$. This (being constant) is the distribution of all n *order statistics* of a random independent sample of size n from $\mathcal{U}(0, t)$. $\square$

The easiest way to generate and display a random realization of a Poisson process is, then, as follows:

```
> (λ, t) := (6.4, 2.0) :
> n := round(Sample (Poisson(λ · t), 1)₁);
```

$$n := 16$$

```
> X := sort(convert (Sample(Uniform(0,t),n), list)) ;
```

$$X := [0.1951, 0.2540, 0.2838, 0.3152, 0.5570, 0.8435, 0.9708, 1.0938, 1.2647,$$

$$1.6006, 1.8116, 1.8268, 1.9143, 1.9150, 1.9298, 1.9412]$$

```
> for i from  1 to  nops(X) do
     cond_i := x < X[i], i − 1;
  end do:
> conditions := seq (cond_i, i = 1..nops(X)) ;
> f := piecewise (conditions, nops(X)) :
```

> $plot\,(f, x = 0..t, numpoints = 500)\,;$

(Output displayed in Fig. 6.1.)

CORRELATION COEFFICIENT

We can easily compute the correlation coefficient between two values of a Poisson process at times t and $t + s$.

We already know the variances are

$$\text{Var}\,(N(t)) = \lambda t,$$
$$\text{Var}\,(N(t + s)) = \lambda(t + s),$$

so all we need is

$$\begin{aligned}
\text{Cov}\,(N(t), N(t + s)) &= \text{Cov}\,(N(t), N(t + s) - N(t) + N(t)) \\
&= \text{Var}\,(N(t)) \\
&= \lambda t
\end{aligned}$$

since $N(t)$ and $N(t + s) - N(t)$ are independent. Clearly, then,

$$\rho_{N(t),N(t+s)} = \frac{\lambda t}{\sqrt{\lambda t} \cdot \sqrt{\lambda(t + s)}} = \frac{1}{\sqrt{1 + \frac{s}{t}}}.$$

The two random variables are thus strongly correlated when s is small and practically uncorrelated when s becomes large, as expected.

Similarly, one can also show the conditional probability of $N(s) = k$, given $N(t) = n$, where $s < t$ (and $k \leq n$), is equal to the following binomial probability:

$$\binom{n}{k} \left(\frac{s}{t}\right)^k \left(1 - \frac{s}{t}\right)^{n-k}.$$

This again relates to the fact that, given $N(t) = n$, the conditional distribution of the n arrival times is uniform over $(0, t)$.

6.2 VARIOUS MODIFICATIONS

There are several ways of extending the Poisson process to deal with more complicated situations. Here are some examples.

SUM OF TWO POISSON PROCESSES

Adding two *independent* Poisson processes with rates λ_1 and λ_2 results in a Poisson process with a rate of $\lambda = \lambda_1 + \lambda_2$ (this follows from the original axioms).

We can also do the opposite: SPLIT a Poisson process into two independent Poisson processes by the following procedure.

A customer stays (buys, registers, etc.) with a probability of p (independently of each other). Then the stream of registered arrivals is a Poisson process, say $X(t)$, with a new rate of $p \cdot \lambda$, and similarly the lost customers constitute a Poisson process, say $Y(t)$, with a rate of $q \cdot \lambda$.

Proposition 6.3. *The two processes $X(t)$ and $Y(t)$ are independent.*

Proof.

$$\Pr\left(X(t) = n \cap Y(t) = m\right)$$
$$= \Pr\left(N(t) = n + m \cap \text{ achieving } n \text{ successes out of } n + m \text{ trials}\right)$$
$$= \frac{(\lambda t)^{n+m}}{(n+m)!} e^{-\lambda t} \cdot \frac{(n+m)!}{n! \cdot m!} p^n q^m$$
$$= \frac{(p\lambda t)^n}{n!} e^{-p\lambda t} \cdot \frac{(q\lambda t)^m}{m!} e^{-q\lambda t}$$
$$= \Pr\left(X(t) = n\right) \cdot \Pr\left(Y(t) = m\right).$$

$\square$

TWO COMPETING POISSON PROCESSES

Example 6.1. Suppose there are two *independent* Poisson processes (such as cars *and* trucks arriving at a gas station) with different (constant) rates λ_1 and λ_2. What is the probability that the first process reaches State n before the second process reaches State m?

Solution. Let $S_n^{(1)}$ and $S_m^{(2)}$ be the corresponding times. We know their distributions are gamma$(n, \frac{1}{\lambda_1})$ and gamma$(m, \frac{1}{\lambda_2})$, respectively. Now, using the following extension of the total probability formula

$$\Pr(A) = \int_L^H \Pr(A \,|\, Y = y) f_Y(y) \, dy$$

we get

$$\Pr\left(S_n^{(1)} < S_m^{(2)}\right) = \int_0^\infty \Pr\left(S_n^{(1)} < S_m^{(2)} \;\middle|\; S_m^{(2)} = t\right) \cdot f_{S_m^{(2)}}(t)\, \mathrm{d}t$$

$$= \int_0^\infty \Pr\left(S_n^{(1)} < t\right) \cdot f_{S_m^{(2)}}(t)\, \mathrm{d}t$$

$$= \int_0^\infty F_{S_n^{(1)}}(t) \cdot f_{S_m^{(2)}}(t)\, \mathrm{d}t$$

$$= \int_0^\infty \left(1 - e^{-t\lambda_1}\left(1 + t\lambda_1 + \frac{(t\lambda_1)^2}{2!} + \cdots + \frac{(t\lambda_1)^{n-1}}{(n-1)!}\right)\right)$$

$$\times\, e^{-t\lambda_2}\frac{(t\lambda_2)^{m-1}}{(m-1)!}\lambda_2\, \mathrm{d}t$$

$$= 1 - \lambda_2^m\left(\frac{1}{(\lambda_1 + \lambda_2)^m} + \frac{\lambda_1 m}{(\lambda_1 + \lambda_2)^{m+1}}\right.$$

$$+ \frac{\lambda_1^2 m(m+1)}{2!(\lambda_1 + \lambda_2)^{m+2}} + \frac{\lambda_1^3 m(m+1)(m+2)}{3!(\lambda_1 + \lambda_2)^{m+3}}$$

$$\left. + \cdots + \frac{\lambda_1^{n-1} m(m+1)(m+2)\cdots(m+n-2)}{(n-1)!(\lambda_1 + \lambda_2)^{m+n-1}}\right).$$

The last result can be rewritten as

$$1 - q\left(q^{m-1} + \binom{m}{1}pq^{m-1} + \binom{m+1}{2}p^2 q^{m-1}\right.$$

$$\left. + \binom{m+2}{3}p^3 q^{m-1} + \cdots + \binom{m+n-2}{n-1}p^{n-1}q^{m-1}\right),$$

where $p = \frac{\lambda_1}{\lambda_1+\lambda_2}$ and $q = 1 - p$. The second term corresponds to the probability of achieving m failures before n successes in a Bernoulli sequence of trials. The same result can also be expressed as the probability of achieving n successes before m failures, namely

$$p\left(p^{n-1} + \binom{n}{1}p^{n-1}q + \binom{n+1}{2}p^{n-1}q^2\right.$$

$$\left. + \binom{n+2}{3}p^{n-1}q^3 + \cdots + \binom{n+m-2}{m-1}p^{n-1}q^{m-1}\right)$$

(a good exercise is to verify that the two answers are identical).

An alternate (and easier) proof of the same formula can be achieved by first not differentiating between trucks and cars and having vehicles arrive at a combined rate of $\lambda_1 + \lambda_2$. With the arrival of each vehicle, we can flip a coin to decide whether it is a car (with a probability of $p = \frac{\lambda_1}{\lambda_1 + \lambda_2}$) or a truck (with a probability of $q = \frac{\lambda_2}{\lambda_1 + \lambda_2}$). The same result then follows immediately.

$\square$

Example 6.2. If cars arrive at a rate of seven per hour, and trucks at a rate of three per hour, what is the probability that the second truck will arrive before the third car?

Solution. If a truck's arrival is considered a success having a probability of $\frac{3}{3+7} = 0.3$, then the chances of two successes happening before three failures are

$$0.3(0.3 + 2 \times 0.3 \times 0.7 + 3 \times 0.3 \times 0.7^2) = 0.3483.$$

Alternatively, we can utilize the corresponding complement (three cars before two trucks):

$$1 - 0.7(0.7^2 + 3 \times 0.7^2 \times 0.3) = 1 - 0.7^3(1 + 3 \times 0.3),$$

which yields the same answer.

$\square$

NONHOMOGENEOUS POISSON PROCESS

When a process is no longer homogeneous (i.e., there are peak and slack periods) and λ is a (known, given) function of t, we must modify the main equation (6.4) to

$$\overset{\bullet}{P}(z,t) = -\lambda(t)(1 - z)P(z,t).$$

Analogously to the $y' = a(x) \cdot y$ equation, whose solution is

$$y(x) = c \cdot \exp\left(\int a(x)\, dx\right),$$

we now get

$$P(z,t) = \exp\left((z - 1)\int_0^t \lambda(s)\, ds\right).$$

The distribution of the number of arrivals during the $(0,t)$ time interval is thus Poisson, with a mean value of $\Lambda_t = \int_0^t \lambda(s)\, ds$.

Note the distribution function for the time of the kth arrival is given by

$$F(t) = 1 - e^{-\Lambda_t} \cdot \sum_{i=0}^{k-1} \frac{\Lambda_t^i}{i!}.$$

This can be extended (by choosing a different time origin) to any other time interval [e.g., the number of arrivals between 10:00 and 11:30 a.m. has a Poisson distribution with a mean of $\int_{10}^{11.5} \lambda(s)\, ds$].

Example 6.3. Assume customers arrive at a rate of 8.4 per hour between 9:00 a.m. and 12:00 p.m.; the rate then jumps to 11.2 during the lunch hour, but starting at 1:00 p.m. it starts decreasing linearly from 11.2 until it reaches 7.3 at 5:00 p.m.. Find the probability of getting more than 25 arrivals between 11:30 a.m. and 2:00 p.m. Also, find the distribution of the third arrival after 1:00 p.m.

Solution.

> $\lambda := t \rightarrow piecewise\left(t < 12, 8.4, t < 13, 11.1, 11.2 - \frac{11.2-7.3}{4} \cdot (t-13)\right):$
> $\lambda(t);$

$$\begin{cases} 8.4 & t < 12 \\ 11.1 & t > 13 \\ 23.8750 - 0.9750\,t & otherwise \end{cases}$$

> $\Lambda := \displaystyle\int_{11.30}^{14} \lambda(t)\, dt;$

$$\Lambda := 27.6925$$

> $1 - \displaystyle\sum_{i=0}^{25} \frac{\Lambda^i}{i!} \cdot e^{-\Lambda};$

$$0.6518$$

> $assume\,(u > 0):$
> $\Lambda := \displaystyle\int_{13}^{13+u} \lambda(t)\, dt :$
> $simplify\,(\Lambda);$

$$\Lambda := 11.2000\,u - .4875\,u^2$$

> $F := 1 - e^{-\Lambda} \cdot simplify\left(\displaystyle\sum_{i=0}^{2} \frac{\Lambda^i}{i!}\right);$

{This is the resulting distribution function (u is the time since 13:00)}

$$F := 1 - e^{(-11.2000\ u + .4875\ u^2)}\ (1.0000 + 11.2000\ u + 62.2325\ u^2$$

$$-5.4600\ u^3 + 0.1188\ u^4)$$

$$> plot\left(\frac{d}{du}F, u = 0..1\right);$$

Time from 13:00

POISSON PROCESS IN MORE DIMENSIONS

The notion of a Poisson process can also be extended to two and three dimensions: the distribution of the number of points (e.g., dandelions, stars) in an area (volume) of size A is Poisson, with the mean of $\lambda \cdot A$, where λ is the point average density. And, given there are exactly n points in a specific region, their conditional distribution is *uniform*.

One can then find (in the three-dimensional case) the distribution of X, the distance from a star to its nearest neighbor, by

$$\Pr(X > x) = \exp\left(-\frac{4}{3}\pi x^3 \lambda\right).$$

This yields the corresponding PDF, namely, $f(x) = 4\pi x^2 \lambda \cdot \exp\left(-\frac{4}{3}\pi x^3 \lambda\right)$, based on which

$$E(X) = \int_0^\infty x f(x)\, dx$$

$$= \left(\frac{4}{3}\lambda\right)^{-\frac{1}{3}} \int_0^\infty u^{\frac{1}{3}} e^{-u}\, du$$

$$= \left(\frac{4}{3}\pi\lambda\right)^{-1/3} \Gamma\left(\frac{4}{3}\right)$$

$$\approx \frac{0.554}{\lambda^{1/3}}.$$

Example 6.4. Consider a two-dimensional Poisson process (of objects we call points) with $\lambda = 13.2$ per unit square, inside a rectangle with opposite corners at $(0,0)$ and $(2,3)$; no points can appear outside this rectangle. Compute the probability of having more than 20 points within 1.2 units of the origin. Also, find the distribution function of the third closest point to the origin.

Solution.

> $\Lambda := \dfrac{\pi}{4} \cdot r^2 \cdot 13.2$:

{area of corresponding quarter-circle, multiplied by the average density}

> $1 - \displaystyle\sum_{i=0}^{20} \dfrac{\Lambda^i}{i!} \cdot e^{-\Lambda} \Bigg|_{r=1.2}$;

$$0.08003$$

> $1 - e^{-\Lambda} \cdot simplify\left(\displaystyle\sum_{i=0}^{2} \dfrac{\Lambda^i}{i!}\right)$;

$$F := 1 - e^{-3.3000\,\pi\,r^2}\left(1.0000 + 10.3673\,r^2 + 53.7400\,r^4\right)$$

> $plot\left(\dfrac{d}{dr}F, r = 0..1.2\right)$;

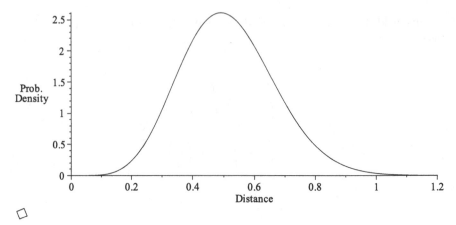

$\square$

$M/G/\infty$ QUEUE

$M/G/\infty$ denotes that the arrivals form a Poisson process (M stands for an older name of the exponential distribution) and are served immediately (there are infinitely many servers), with the service time being a random variable having a distribution function $G(x)$; the individual service times are also *independent* of each other (let us call them S_1, S_2, ...).

If $X(t)$ is the number of customers in a SYSTEM (i.e., being served), partitioning the sample space according to how many have arrived during the $[0, t]$ interval, we get

$$\Pr\left(X(t) = j\right) = \sum_{n=0}^{\infty} \Pr\left(X(t) = j \mid N(t) = n\right) \cdot \frac{(\lambda t)^n}{n!} e^{-\lambda t}.$$

Given $N(t) = n$, the n arrivals are distributed *uniformly* over $[0, t]$. Given a customer arrived at time x, the probability of his still being served at time t is

$$\Pr(S > t - x) = 1 - G(t - x).$$

Since x itself has a uniform distribution over $[0, t]$, the probability of his departure time, say $T > t$ (which means at time t he is still being served), is computed by

$$p_t = \int_0^t \Pr(T > t \mid x) \cdot f(x)\, \mathrm{d}x$$

$$= \int_0^t \Pr(S > t - x) \frac{\mathrm{d}x}{t}$$

$$= \int_0^t (1 - G(t-x)) \frac{dx}{t}$$

$$= \int_0^t (1 - G(u)) \frac{du}{t}.$$

The final answer is therefore

$$\Pr(X(t) = j) = \sum_{n=j}^{\infty} \binom{n}{j} p_t^j q_t^{n-j} \cdot \frac{(\lambda t)^n}{n!} e^{-\lambda t}$$

$$= e^{-\lambda t} \frac{(\lambda t p_t)^j}{j!} \sum_{n=j}^{\infty} \frac{(\lambda t q_t)^{n-j}}{(n-j)!}$$

$$= e^{-\lambda t} \frac{(\lambda t p_t)^j}{j!} e^{\lambda t q_t}$$

$$= \frac{(\lambda t p_t)^j}{j!} e^{-\lambda t p_t},$$

which is a Poisson distribution with a mean of

$$\Lambda_x = \lambda t p_t = \lambda \int_0^t (1 - G(u)) \, du.$$

Investigating the stationary distribution of the process as $t \to \infty$ we first obtain

$$\int_0^t (1 - G(u)) \, du \xrightarrow{t \to \infty} \int_0^{\infty} (1 - G(u)) \, du = \int_0^{\infty} u' (1 - G(u)) \, du$$

$$= \int_0^{\infty} u f(u) \, du,$$

that is, the average service time. In this limit, Λ_x is thus the ratio of the average service time to the average interarrival time.

Note the resulting Poisson probabilities also represent the *proportion* of time spent in each state in the *long run*.

Let us now introduce $Y(t)$ for the number of customers who, by the time t, have already been served and left the system. Are $X(t)$ and $Y(t)$ independent? Let us see:

$$\Pr\left(X(t) = j \cap Y(t) = i\right)$$

$$= \sum_{n=0}^{\infty} \Pr\left(X(t) = j \cap Y(t) = i \mid N(t) = n\right) \cdot \frac{(\lambda t)^n}{n!} e^{-\lambda t}$$

$$= \Pr\left(X(t) = j \cap Y(t) = i \mid N(t) = i + j\right) \cdot \frac{(\lambda t)^{i+j}}{(i+j)!} e^{-\lambda t}$$

$$= \binom{i+j}{j} p_t^j q_t^i \cdot \frac{(\lambda t)^{i+j}}{(i+j)!} e^{-\lambda t}$$

$$= \frac{(\lambda t p_t)^j}{j!} e^{-\lambda t p_t} \cdot \frac{(\lambda t q_t)^i}{i!} e^{-\lambda t q_t}.$$

The answer is YES, and individually all $X(t)$ and $Y(t)$ have a Poisson distribution with a mean of $\lambda t p_t$ and $\lambda t q_t$, respectively.

Notice this does *not* imply that the two processes are Poisson – neither of them is! But it *does* give us the means of finding the distribution of the time of the, say, third *departure*.

Example 6.5. Consider an $M/G/\infty$ queue with customers arriving at a rate of 48 per hour and service times having a gamma$(7, 3\,\text{min})$ distribution. If we start the process with no customers, what is the probability that, 25 min later, there are at least 15 customers being serviced while more than 5 have already left.

Also, find the expected time and standard deviation of the time of the second departure.

Solution.

{This is the PDF of the gamma(7,3) distribution,}

$$> g := \frac{t^6 \cdot e^{-\frac{t}{3}}}{6! \cdot 3^7} :$$

$$> G := \int_0^u g \; dt :$$

$$> (t, \lambda) := \left(25., \frac{48}{60}\right) \;\; \{\text{rate is per minute, to be consistent}\}$$

$$> p_t := \frac{1}{t} \cdot \int_0^t (1 - G) \; du;$$

$$p_{25.00} := 0.7721$$

$$> \Lambda_x := \lambda \cdot t \cdot p_t;$$

$$\Lambda_x := 15.4419$$

> $\Lambda_y := \lambda \cdot t \cdot (1 - p_t);$

$$\Lambda_y := 4.5581$$

> $\left(1 - \sum_{i=0}^{14} \frac{\Lambda_x^i}{i!} \cdot e^{-\Lambda_x}\right) \cdot \left(1 - \sum_{i=0}^{5} \frac{\Lambda_y^i}{i!} \cdot e^{-\Lambda_y}\right);$

$$0.1777$$

> $t := evaln(t) :\{$release the value of $t.\}$

> $\Lambda := \lambda \cdot \int_0^t G \, du:$

> $F := 1 - (1 + \Lambda) \cdot e^{-\Lambda}:$

> $\mu := evalf \left(\int_0^\infty t \cdot \frac{dF}{dt} dt\right);$

$$19.0853$$

> $\sqrt{evalf \left(\int_0^\infty (x - \mu)^2\right) \cdot \frac{dF}{dt} \, dt};$

$$3.6363$$

$\square$

COMPOUND (CLUSTER) POISSON PROCESS

Assume a Poisson process represents arriving customers and that the jth customer will make a random purchase of amount Y_j (these are independent and identically distributed); alternatively, customers may arrive in *groups* of size Y_j (ignore how much they buy), which explains why it is called a CLUSTER. Using the first interpretation of Y_j, we are interested in the total amount of money spent by those customers who arrived during the time interval $(0, t)$, or

$$Y(t) \equiv \sum_{j=1}^{N(t)} Y_j.$$

The moment-generating function (MGF) of $Y(t)$ is thus

$$\mathbb{E}\left(e^{u \cdot Y(t)}\right) = \sum_{n=0}^{\infty} \mathbb{E}\left(e^{u \cdot Y(t)} \mid N(t) = n\right) \frac{\lambda^n t^n}{n!} e^{-\lambda t}$$

$$= e^{-\lambda t} \sum_{n=0}^{\infty} \mathbb{E}\left(\exp\left(u \sum_{j=1}^{n} Y_j\right)\right) \frac{\lambda^n t^n}{n!}$$

$$= e^{-\lambda t} \sum_{n=0}^{\infty} M_Y(u)^n \frac{\lambda^n t^n}{n!}$$

$$= \exp\left(-\lambda t \left(1 - M_Y(u)\right)\right),$$

where $M_Y(u)$ is the MGF of each single purchase Y_j.

The expected value of $Y(t)$ is simply $\lambda t \mu_Y$ (just differentiate the preceding expression with respect to u and evaluate at $u = 0$).

Proposition 6.4.

$$Var\left(Y(t)\right) = \lambda t \mathbb{E}(Y_i^2) = \lambda t \sigma_Y^2 + \lambda t \mu_Y^2.$$

Proof. The second simple moment of $Y(t)$ is

$$\frac{d^2}{du^2} \exp\left(-\lambda t \left(1 - M_Y(u)\right)\right)\Big|_{u=0}$$

$$= \left(\lambda t M_Y''(u) + \lambda^2 t^2 M_Y'(u)^2\right) \exp\left(-\lambda t \left(1 - M_Y(u)\right)\right)\Big|_{u=0}$$

$$= \lambda t \left(\sigma_Y^2 + \mu_Y^2\right) + \lambda^2 t^2 \mu_Y^2,$$

from which we subtract $(\lambda t \mu_Y)^2$ to get the variance.   □

The first (second) term represents the variation due to the random purchases (random number of arrivals).

When Y_n is of the integer (cluster) type, we can do even better: find the PGF of $Y(t)$:

$$\exp\left(-\lambda t \left(1 - P(z)\right)\right),$$

where $P(z)$ is the PGF of each Y_j.

Example 6.6. Suppose customers arrive in clusters, at an average rate of 26 clusters per hour. The size of each cluster is (independently of the other clusters) a random variable with $P(z) = z \cdot (0.8 + 0.2z)^5$. Find and display the distribution of the total number of customers who arrive during the next 15 min and the corresponding mean and standard deviation.

Solution.

> $P := z \cdot (0.8 + 0.2 \cdot z)^5$:
> $PGF := e^{\lambda \cdot t \cdot (P-1)}$;

$$PGF := e^{\lambda t \left(z(0.8+0.2z)^5 - 1 \right)}$$

> $(t, \lambda) := \left(\dfrac{15}{60}, 26 \right)$: {this time we use hours as units of time}
> $prob := mtaylor(PGF, z, 35)$:
> $pointplot([seq([i, coeff(prob, z, i)], i = 0..34)])$

> $\mu := \left. \dfrac{d}{dz} PGF \right|_{z=1}$;

$$\mu := 13.0000$$

> $\sigma := \sqrt{\left. \dfrac{d^2}{dz^2} PGF \right|_{z=1} + \mu - \mu^2}$;

$$\sigma := 5.5857$$

$\square$

POISSON PROCESS OF RANDOM DURATION

Suppose a Poisson process is terminated at a random time T. We would like to get the mean and variance of $N(T)$, which is the total number of arrivals.

Using the formula of total expected value, we get

$$\mathbb{E}\left(N(T)\right) = \int_0^\infty \mathbb{E}\left(N(T)\,|\,T=t\right)\cdot f(t)\,\mathrm{d}t = \int_0^\infty \lambda t\cdot f(t)\,\mathrm{d}t = \lambda\cdot\mathbb{E}(T)$$

(a rather natural result). Similarly,

$$\mathbb{E}\left(N(T)^2\right) = \int_0^\infty \mathbb{E}\left(N(T)^2\,|\,T=t\right)\cdot f(t)\,\mathrm{d}t$$

$$= \int_0^\infty \left(\lambda t + \lambda^2 t^2\right)\cdot f(t)\,\mathrm{d}t$$

$$= \lambda\mathbb{E}(T) + \lambda^2\cdot\left(\mathrm{Var}(T) + \mathbb{E}(T)^2\right),$$

which implies

$$\mathrm{Var}\left(N(T)\right) = \lambda\mathbb{E}(T) + \lambda^2\mathrm{Var}(T).$$

The first term reflects the variance in the random number of arrivals, and the second one is the contribution due to random T.

It is not too difficult to show the PGF of $N(T)$ is

$$\mathbb{E}\left(z^{N(T)}\right) = \int_0^\infty \mathbb{E}\left(z^{N(T)}\,\Big|\,T=t\right)\cdot f(t)\,\mathrm{d}t$$

$$= \int_0^\infty e^{\lambda(z-1)t}\cdot f(t)\,\mathrm{d}t$$

$$= M\left((z-1)\lambda\right),$$

where $M(u)$ is the *moment*-generating function of T.

Example 6.7. A Poisson process with $\lambda = 4.9$ per hour is observed for a random time T, distributed uniformly between 5 and 11 h. Find and display the distribution of the total number of arrivals thus observed, and compute the corresponding mean and standard deviation.

Solution.

{MGF of uniform distribution – see Sect. 12.2.}

> $M := u \to \dfrac{e^{11\cdot u} - e^{5\cdot u}}{6\cdot u}$:

> $PGF := M\left(\lambda\cdot(z-1)\right)$;

$$PGF := \frac{1}{6}\,\frac{e^{11\,\lambda(z-1)} - e^{5\,\lambda(z-1)}}{\lambda(z-1)}$$

> $\lambda := 4.9$:

```
> prob := mtaylor(PGF, z, 80) :
> pointplot ([seq ([i, coeff (prob, z, i)], i = 10..79)]) ;
```

Number of arrivals

$$> \mu := \lim_{z \to 1} \left(\frac{d}{dz} PGF \right);$$

$$\mu := 39.200$$

{or, alternatively, by}

$$> \frac{11 + 5}{2} \cdot \lambda;$$

$$39.2000$$

$$> \sigma := \sqrt{\lim_{z \to 1} \left(\frac{d^2}{dz^2} PGF \right) + \mu - \mu^2}; \ \{\text{based on PGF:}\}$$

$$\sigma := 10.5466$$

$$> \sqrt{\frac{(11 + 5)}{2} \cdot \lambda + \frac{(11 - 5)^2}{12} \cdot \lambda^2}; \ \{\text{using our formula:}\}$$

$$10.5466$$

$\square$

EXERCISES

Exercise 6.1. Consider a Poisson process with a rate function given by $\lambda(t) = 1 + \frac{\sin t}{2}$. Calculate the probability of more than three arrivals during an interval of $0.5 < t < 1.2$. Find the distribution function of the time of the second arrival and the corresponding mean and standard deviation.

Exercise 6.2. Customers arrive at a rate given by the following expression: $\lambda(t) = 2.7e^{-t/3}$. Find:

(a) The probability of fewer than 5 arrivals between $t = 1$ and $t = 2$;
(b) The correlation coefficient between the number of arrivals in the $(0, 1)$ time interval and in the $(0, 2)$ time interval;
(c) The distribution function $F(t)$ of the time of the third arrival and its value at $t \to \infty$. Why is the limit less than 1?

Exercise 6.3. Suppose that customers arrive at a rate of 14.7 clusters per hour, where the size of each cluster has the following distribution:

Cluster size	1	2	3	4
Pr	0.36	0.32	0.18	0.14

Find:

(a) The expected number of *customers* to arrive during the next 42 min and the corresponding standard deviation;
(b) The probability that the number of *customers* who arrive during the next 42 min will be between 14 and 29 (inclusive);
(c) The probability that at least one of the *clusters* arriving during the next 42 min will be greater than 3. *Hint*: derive the answer using the total probability formula.

Exercise 6.4. Consider an $M/G/\infty$ queue with service times having a gamma(2, 13 min) distribution and customers arriving at a rate of 12.6 per hour. Find:

(a) The probability that, 17 min after the store opens (with no customers waiting at the door), there will be exactly 2 busy servers;
(b) The long-run average of busy servers;
(c) The probability that the service times of the first 3 customers will all be shorter than 20 min (each).

Exercise 6.5. Customers arrive at a constant rate of 14.7 per hour, but each of them will make a purchase (instantly, we assume, and independently of each other) only with a probability of 67% (otherwise, they will only browse). The value of a single purchase is, to a good approximation, a random variable having a gamma(2, $13) distribution.

(a) Compute the probability that, during the next 40 min, the store will get at least 12 customers who buy something and (at the same time – this is a single question) no more than 7 who will not make any purchase;
(b) Compute the probability that, by the time the store gets its ninth buying customer, it will have had no more than five browsing ones;
(c) Find the expected value of the total purchases made during the next 40 min and the corresponding standard deviation.

Exercise 6.6. Consider a three-dimensional Poisson process with $\lambda = 193$ dots per cubic meter. Find the expected value and standard deviation of:

(a) The number of dots in the region defined by $x^2 + y^2 < 0.37$ and (at the same time) $0 < z < 1$;
(b) The distance from a dot to its nearest neighbor;
(c) The distance from a dot to its fifth nearest neighbor.

Exercise 6.7. A Poisson process with an arrival rate of 12.4 per hour is observed for a random time T whose distribution is gamma$(5, 12\,\text{min})$. Compute:

(a) The expected value and standard deviation of the total number of arrivals recorded;
(b) The probability that this number will be between 10 and 20 (inclusive);
(c) $\Pr(T > 80\,\text{min})$.

CHAPTER 7
Birth and Death Processes I

We generalize the Poisson process in two ways:

1. By letting the value of the arrival rate λ depend on the current state n;
2. By including DEPARTURES, which allows the process to instantaneously *decrease* its value by one unit (at a rate that will also be a function of n).

These generalizations can then be used to describe not only customers entering *and* leaving a store, but also populations that increase or decrease in size due to the birth of a new member or the death of an old one.

7.1 BASICS

We now investigate the case where the process (currently in state n) can either go up one step (this happens at a rate of λ_n) or go down one step (at a rate of μ_n). Note both λ_n and μ_n are now (nonnegative) functions of n, with the single restriction of $\mu_0 = 0$ (the process cannot enter negative values).

Using the approach presented in the previous chapter, one can show this leads to the following set of difference-differential equations:

$$\dot{P}_{i,n}(t) = -(\lambda_n + \mu_n)P_{i,n}(t) + \lambda_{n-1}P_{i,n-1}(t) + \mu_{n+1}P_{i,n+1}(t), \qquad (7.1)$$

which can be solved in only several special cases (discussed in individual sections of this chapter).

A way to understand the preceding equations is to realize the probability of being in state n can change in one of three ways: if we are currently in state n, we will leave it at a rate of $\lambda_n + \mu_n$ (this will *decrease* the probability of being in state n – thus the minus sign); if we are in state $n-1$, we will

J. Vrbik and P. Vrbik, *Informal Introduction to Stochastic Processes with Maple*, Universitext, DOI 10.1007/978-1-4614-4057-4_7,
© Springer Science+Business Media, LLC 2013

enter state n at a rate λ_{n-1}; if we are in state $n + 1$, we will enter state n at a rate of μ_{n+1}.

To simulate a random realization of any such process (starting in state i), we first generate the time till the first TRANSITION (either up or down), which is exponentially distributed with a mean of $\frac{1}{\lambda_i + \mu_i}$ (based on the *combined* rate of a transition happening during an infinitesimal time interval).

Given this first transition happened during $[t, t + \Delta)$, the conditional probability of going up one step is

$$\frac{e^{-(\lambda_i + \mu_i)} (\lambda_i \Delta + o(\Delta))}{e^{-(\lambda_i + \mu_i)} ((\lambda_i + \mu_i)\Delta + o(\Delta))} \xrightarrow{\Delta \to \infty} \frac{\lambda_i}{\lambda_i + \mu_i}$$

($\frac{\mu_i}{\lambda_i + \mu_i}$ is the corresponding probability of going down). We can thus easily decide (based on a random flip of a correspondingly biased coin) which way to move. This procedure can be repeated with the new value of i as many times as needed.

We mention in passing that, alternatively (but equally correctly), we may generate the tentative time of the next move up (using an exponential distribution with a mean of $\frac{1}{\lambda_i}$) and of the next move down (exponential, with a mean of $\frac{1}{\mu_i}$) and let them compete (i.e., take the one that happens earlier; the other one must then be *discarded* because the process no longer continues in state i). This must then be repeated with the *new* rates of the state just entered (and *new* tentative up and down moves, of which *only* the earlier one is actually taken). One can show this procedure is probabilistically equivalent to (but somehow more clumsy than) the previous one.

We use the following program to visualize what a development of any such process looks like:

> $\lambda := n \to 15 \cdot 0.8^n$:

{Define your birth rates}

> $\mu := n \to \dfrac{6 \cdot n}{1 + 0.3 \cdot n}$:

{and your death rates.}

> $(i, T) := (3, 2)$:

{Specify initial state and final time T.}

> $(tt, n) := (0, i)$:

{Initialize auxiliary variables.}

> **for** j **while** $tt < T$ **and** $\lambda(n) + \mu(n) > 0$ **do**

$$tt := tt + Sample\left(Exponential\left(\frac{1}{\lambda(n) + \mu(n)}\right), 1\right)_1 :$$

$$n := n + \begin{cases} 1 & Sample\,(Uniform(0, 1), 1)_1 < \dfrac{\lambda(n)}{\lambda(n) + \mu(n)} \\[2mm] -1 & otherwise \end{cases} :$$

> $cond_j := t < tt, n :$
> **end do:**
> $conditions := seq\left(cond_j, j = 1..j - 1\right) :$
> $f := piecewise(conditions, t = 0..T, numpoints = 500);$

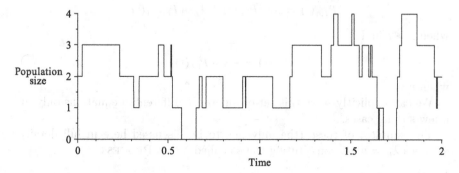

The set of equations (7.1) is impossible to solve analytically unless a specific and simple choice of λ_n and μ_n is made. In subsequent sections we investigate many such special models.

7.2 PURE-BIRTH PROCESS

Consider an extension of the Poisson process, where the rate at which the process jumps to the next (higher) state depends on n (the current state), but there are no deaths.

First, we define $P_{i,n}(t)$ as

$$P_{i,n}(t) \equiv \Pr\left(X(t) = n \mid X(0) = i\right).$$

The process is still homogeneous in time, that is,

$$\Pr\left(X(t + s) = n \mid X(t) = i\right) = P_{i,n}(s)$$

for any $t > 0$ and $s > 0$, but now

$$\Pr\left(X(t + s) - X(t) = 1 \mid X(t) = n\right) = P_{n,n+1}(s) = \lambda_n \cdot s + o(s),$$

$$\Pr\left(X(t + s) - X(t) \geq 2 \mid X(t) = n\right) = \sum_{j=2}^{\infty} P_{n,n+j}(s) = o(s),$$

which implies

$$\Pr\left(X(t + s) - X(t) = 0 \mid X(t) = n\right) = P_{n,n}(s) = 1 - \lambda_{\cdot n} \cdot s + o(s).$$

Based on the formula of total probability, we now get

$$P_{i,n}(t+s) = P_{i,n}(t)P_{n,n}(s) + P_{i,n-1}(t)P_{n-1,n}(s)$$
$$+ P_{i,n-2}(t)P_{n-2,n}(s) + \cdots + P_{i,i}(t)P_{i,n}(s).$$

Subtracting $P_{i,n}(t)$, dividing by s, and taking the $s \to 0$ limit yields

$$\overset{\bullet}{P}_{i,n}(t) = -\lambda_n P_{i,n}(t) + \lambda_{n-1} P_{i,n-1}(t)$$

when $n > i$ and

$$\overset{\bullet}{P}_{i,i}(t) = -\lambda_i \cdot P_{i,i}(t)$$

when $n = i$.

We can explicitly solve this set of difference-differential equations only in a few special cases.

The simplest of these (the only one to be discussed here in full detail) assumes $\lambda_n = n \cdot \lambda$, constituting the so-called YULE PROCESS.

YULE PROCESS

With the help of the PGF idea, that is,

$$P_i(z,t) = \sum_{n=i}^{\infty} P_{i,n}(t) \cdot z^n,$$

we get

$$\overset{\bullet}{P}_i(z,t) = -\lambda \sum_{n=i}^{\infty} nz^n P_{i,n}(t) + \lambda \sum_{n=i}^{\infty} nz^{n+1} P_{i,n}(t)$$
$$= -\lambda z P_i'(z,t) + \lambda z^2 P_i'(z,t)$$
$$= -\lambda z (1-z) P_i'(z,t),$$

where the prime indicates differentiation with respect to z. We are thus faced with solving a simple PARTIAL DIFFERENTIAL EQUATION (PDE) in two variables, t and z. We also know $P(z,0) = z^i$ (the initial-value condition).

The complete solution is[1]

$$P_i(z,t) = \left(\frac{z\,p_t}{1 - z q_t} \right)^i,$$

where $p_t = \mathrm{e}^{-\lambda t}$ and $q_t = 1 - \mathrm{e}^{-\lambda t}$.

---

[1] For further instruction on how to solve such PDEs see Appendix 7.A.

Do we recognize the corresponding distribution? Yes, it is the negative binomial distribution (waiting for the ith success), where the probability of a success, namely p_t, depends on time. This implies

$$\mathbb{E}\left(X(t)\right) = \frac{i}{p_t} = i \cdot e^{\lambda t}$$

(exponential population explosion) and

$$\text{Var}\left(X(t)\right) = \frac{i}{p_t}\left(\frac{1}{p_t} - 1\right) = i \cdot e^{\lambda t} \cdot (e^{\lambda t} - 1).$$

Example 7.1. Suppose $\lambda = 3.2/\min$ and the Yule process starts in State 3. What is the probability that, 45 s later, it will have exactly eight members? More than ten members? Also, compute the expected value and standard deviation of $X(45\,\text{s})$.

Solution. $p = e^{-3.2 \times \frac{3}{4}} = 0.090718$ (note we are using minutes as our units of time). We will need exactly eight trials to achieve the third success with a probability of

$$\binom{7}{2} p^2 q^5 \cdot p = 0.9745\%.$$

We will need more than ten trials with a probability of

$$q^{10} + 10 p q^9 + \binom{10}{2} p^2 q^8 = 94.487$$

(think of what must happen during the first ten trials). The expected value of $X(45\,\text{s})$ is $\frac{3}{p} = 33.0695$, and the standard deviation equals $\sqrt{\frac{3}{p}\left(\frac{1}{p} - 1\right)} =$ 18.206.

$\square$

To find the probability of exactly j births (i.e., the value of the process increasing by j) between time t and $t+s$, we break down the answer according to the state reached at time t and use the total-probability formula (and the Markovian property) thus:

$$\Pr\left(X(t+s) - X(t) = j \mid X(0) = i\right)$$

$$= \sum_{k=i}^{\infty} \Pr\left(X(t+s) = j + k \mid X(t) = k\right) \cdot \Pr\left(X(t) = k \mid X(0) = i\right).$$

This results in

$$\sum_{k=i}^{\infty} \binom{j+k-1}{k-1} e^{-\lambda s k} (1 - e^{-\lambda s})^j \cdot \binom{k-1}{i-1} e^{-\lambda t i} (1 - e^{-\lambda t})^{k-i}$$

$$= \binom{j+i-1}{i-1} \cdot (1 - e^{-\lambda s})^j \cdot e^{-\lambda t i} \sum_{k=i}^{\infty} \binom{j+k-1}{k-i} e^{-\lambda s k} (1 - e^{-\lambda t})^{k-i}$$

$$= \binom{j+i-1}{i-1} \cdot (1 - e^{-\lambda s})^j \cdot e^{-\lambda t i} \sum_{m=0}^{\infty} \binom{m+i+j-1}{m} e^{-\lambda s(m+i)} (1 - e^{-\lambda t})^m$$

$$= \binom{j+i-1}{i-1} \cdot (1 - e^{-\lambda s})^j \cdot e^{-\lambda t i} \cdot e^{-\lambda s i} \sum_{m=0}^{\infty} \binom{-i-j}{m} (-1)^m e^{-\lambda s m} (1 - e^{-\lambda t})^m$$

$$= \binom{j+i-1}{i-1} \left(\frac{e^{-\lambda(t+s)}}{1 - e^{-\lambda s} + e^{-\lambda(t+s)}} \right)^i \left(\frac{1 - e^{-\lambda s}}{1 - e^{-\lambda s} + e^{-\lambda(t+s)}} \right)^j$$

for $j = 0, 1, 2, \ldots$. This can be identified as the *modified* negative binomial distribution (counting only the number of *failures* till the ith success), with the probability of success given by

$$\frac{e^{-\lambda(t+s)}}{1 - e^{-\lambda s} + e^{-\lambda(t+s)}}$$

whose value is always between 0 and 1.

7.3 PURE-DEATH PROCESS

The basic assumptions are now that the process can only *lose* its members, according to

$$\Pr\left(X(t + s) - X(t) = -1 \mid X(t) = n\right) = P_{n,n-1}(s) = \mu_n \cdot s + o(s)$$

$$\Pr\left(X(t + s) - X(t) \geq -2 \mid X(t) = n\right) = \sum_{j=2}^{\infty} P_{n,n-j}(s) = o(s),$$

implying

$$\Pr\left(X(t + s) - X(t) = 0 \mid X(t) = n\right) \equiv P_{n,n}(s) = 1 - \mu_n \cdot s + o(s),$$

where μ_0 must be equal to zero (State 0 is thus absorbing).

These translate to

$$\dot{P}_{i,n}(t) = -\mu_n P_{i,n}(t) + \mu_{n+1} P_{i,n+1}(t).$$

We will solve this set of difference-differential equations only in the special case of

$$\mu_n = n \cdot \mu.$$

Multiplying by z^n and summing over n from 0 to i (the only possible states now) yields

$$\dot{P}_i(z,t) = -\mu z \, P_i'(z,t) + \mu P_i'(z,t) = \mu(1-z) P_i'(z,t),$$

where

$$P_i(z,t) = \sum_{n=0}^{i} P_{i,n}(t) \cdot z^n.$$

Solving the PDE (Appendix 7.A) with the usual initial condition of $P(z,0) = z^i$, we get

$$P_i(z,t) = (q_t + p_t z)^i,$$

where $p_t = e^{-\mu t}$. The resulting distribution is binomial, with a mean value of $i \cdot e^{-\mu t}$ and variance of $i \cdot e^{-\mu t} \cdot (1 - e^{-\mu t})$.

In addition to questions like those from the last example, we may also want to investigate TIME TILL EXTINCTION, say T. Clearly, $\Pr(T \le t) = P_{i,0}(t) = (1 - e^{-\mu t})^i$. We thus get

$$\mathbb{E}(T) = \int_0^\infty t \cdot \frac{d\left((1 - e^{-\mu t})^i - 1\right)}{dt} \, dt$$

$$= \int_0^\infty \left(1 - (1 - e^{-\mu t})^i\right) \, dt$$

$$= \int_0^\infty \left(\sum_{j=1}^{i} (-1)^{j+1} \binom{i}{j} e^{-j\mu t} \right) dt$$

$$= \frac{1}{\mu} \sum_{j=1}^{i} \frac{(-1)^{j+1}}{j} \binom{i}{j},$$

or, more easily,

$$\mathbb{E}(T) = \frac{1}{\mu} \left(1 + \frac{1}{2} + \frac{1}{3} + \cdots + \frac{1}{i} \right)$$

since the distribution of time till the next transition is *exponential* with a mean value of $\frac{1}{n\mu}$, given we are currently in State n. (Why are the two results different? Well, since they are both correct, there can only be one explanation: the two formulas are equivalent). Furthermore, these times are *independent* of each other, which enables us to also find

$$\text{Var}(T) = \frac{1}{\mu^2}\left(1 + \frac{1}{2^2} + \frac{1}{3^2} + \cdots + \frac{1}{i^2}\right).$$

7.4 LINEAR-GROWTH MODEL

We combine the previous model with the Yule process; thus,

$$\lambda_n = n \cdot \lambda$$

and

$$\mu_n = n \cdot \mu.$$

This leads to

$$\begin{aligned}
\dot{P}_i(z,t) &= -(\lambda + \mu)z\,P_i'(z,t) + \lambda z^2 P_i'(z,t) + \mu P_i'(z,t) \\
&= (\mu - \lambda z)(1 - z)P_i'(z,t),
\end{aligned}$$

whose solution is (Appendix 7.A)

$$P_i(z,t) = \left(r_t + (1 - r_t)\frac{p_t z}{1 - q_t z}\right)^i,$$

where

$$r_t = \frac{\mu(1 - e^{-(\lambda-\mu)t})}{\lambda - \mu e^{-(\lambda-\mu)t}} \qquad\qquad 1 - r_t = \frac{\lambda - \mu}{\lambda - \mu e^{-(\lambda-\mu)t}}$$

$$p_t = \frac{(\lambda - \mu)e^{-(\lambda-\mu)t}}{\lambda - \mu e^{-(\lambda-\mu)t}} \qquad\qquad q_t = \frac{\lambda(1 - e^{-(\lambda-\mu)t})}{\lambda - \mu e^{-(\lambda-\mu)t}}$$

(one can and should verify all these are probabilities between 0 and 1).

Note when $\lambda = \mu$, based on L'Hopital's rule, we get (differentiating with respect to λ, then setting $\lambda = \mu$):

$$p_t = \frac{1}{1 + \mu t}$$

$$r_t = q_t = \frac{\mu t}{1 + \mu t}.$$

When expanded, the PGF (a composition of binomial and geometric distributions) yields the following explicit formulas for individual and cumulative probabilities of the $X(t)$ distribution:

$$\Pr(X(t) = 0 \mid X(0) = i) = r_t^i,$$

$$\Pr(X(t) = k \mid X(0) = i),$$

$$= \sum_{j=1}^{\min(i,k)} \binom{i}{j}(1 - r_t)^j r_t^{i-j} \cdot p_t^j \binom{k-1}{j-1}(1 - p_t)^{k-j} \quad \text{when } k > 0,$$

$$\Pr(X(t) \le \ell \mid X(0) = i)$$

$$= r_t^i + \sum_{j=1}^{\min(i,\ell)} \binom{i}{j}(1 - r_t)^j r_t^{i-j} \cdot p_t^j \sum_{k=j}^{\ell} \binom{k-1}{j-1}(1 - p_t)^{k-j}.$$

MEAN AND STANDARD DEVIATION

As $P(z,t)$ is a COMPOSITION of two PGFs, $G(z) = (r_t + (1 - r_t)z)^i$ of a usual binomial distribution (i is the number of trials, $1 - r_t$ the probability of a success) and $F(z) = \frac{p_t z}{1 - q_t z}$ of a geometric distribution, we can find the corresponding expected value and standard deviation of $X(t)$ as follows.

By differentiating $G(F(z))$, we get $G'(F(z)) \cdot F'(z)$, which implies the composite mean is simply the product of the individual means, say m_1 and m_2. In our case, this yields

$$\mathbb{E}(X(t)) = i(1 - r_t) \cdot \frac{1}{p_t} = i \cdot e^{(\lambda - \mu)t}.$$

Differentiating one more time with respect to z results in $G''(F')^2 + G'F''$. Converting to the corresponding variance yields

$$(V_1 - m_1 + m_1^2)m_2^2 + (V_2 - m_2 + m_2^2)m_1 + m_1 m_2 - m_1^2 m_2^2 = V_1 m_2^2 + V_2 m_1.$$

In this case, we get

$$\text{Var}(X(t)) = i\frac{r_t(1 - r_t)}{p_t^2} + i\frac{q_t(1 - r_t)}{p_t^2} = i\frac{\lambda + \mu}{\lambda - \mu}\left(e^{(\lambda - \mu)t} - 1\right)e^{(\lambda - \mu)t}.$$

EXTINCTION

The probability of being extinct at time t is equal to r_t^i; ULTIMATE EXTINCTION has a probability of

$$\lim_{t \to \infty} r_t^i = \begin{cases} \left(\frac{\mu}{\lambda}\right)^i & \lambda > \mu, \\ 1 & \lambda \le \mu. \end{cases}$$

Mean Time Till Extinction

When extinction is certain ($\lambda \leq \mu$), we get, based on the previous set of formulas, the following distribution function of the random time (say T) till extinction:

$$\Pr(T \leq t) = r_t^i = \left(\frac{\mu(1 - e^{-(\lambda-\mu)t})}{\lambda - \mu e^{-(\lambda-\mu)t}} \right)^i .$$

To get the corresponding expected value of T, we start with the case of $i = 1$.

$$\mathbb{E}(T \mid X(0) = 1) = \int_0^\infty t \cdot \frac{d(r_t - 1)}{dt} \, dt$$

$$= \int_0^\infty (1 - r_t) \, dt$$

$$= \int_0^\infty \frac{\lambda - \mu}{\lambda - \mu e^{-(\lambda-\mu)t}} \, dt$$

$$= \int_0^\infty \frac{(\lambda - \mu)e^{(\lambda-\mu)t}}{\lambda e^{(\lambda-\mu)t} - \mu} \, dt$$

$$= \int_0^1 \frac{dx}{\mu - \lambda x}$$

$$= -\frac{1}{\lambda} \ln(\mu - \lambda x) \Big|_{x=0}^1$$

$$= \frac{1}{\lambda} \ln \frac{\mu}{\mu - \lambda} .$$

To extend this to an arbitrary i, we define $\omega_i = \mathbb{E}(T \mid X(0) = i)$ for every nonnegative integer i. These must follow the following set of difference equations:

$$\omega_i = \frac{\mu}{\mu + \lambda} \omega_{i-1} + \frac{\lambda}{\mu + \lambda} \omega_{i+1} + \frac{1}{i(\mu + \lambda)},$$

where $i(\mu + \lambda)$ is the overall rate for making a transition (the reciprocal yields the expected value of the time till it happens), $\frac{i\mu}{i\mu+i\lambda} = \frac{\mu}{\mu+\lambda}$ is the conditional probability of the corresponding jump taking the process one step down, and $\frac{\lambda}{\mu+\lambda}$ is the conditional probability of taking it one step up.

Even though we do not know how to solve this set of difference equations analytically (the nonhomogeneous term is *not* a polynomial in i), we can solve it recursively, knowing the value of $\omega_0 = 0$ and of $\omega_1 = \frac{1}{\lambda} \ln \frac{\mu}{\mu - \lambda}$. Thus, we get

$$\omega_2 = \frac{\mu + \lambda}{\lambda^2} \ln \frac{\mu}{\mu - \lambda} - \frac{1}{\lambda},$$

$$\omega_3 = \frac{\mu^2 + \mu\lambda + \lambda^2}{\lambda^3} \ln \frac{\mu}{\mu - \lambda} - \frac{2\mu + 3\lambda}{2\lambda^2},$$

etc.

Continuing this sequence with the help of MAPLE:

```
> w0 := 0 :
```

$$> w_1 := \frac{\ln\left(\frac{\mu}{\mu - \lambda}\right)}{\lambda} :$$

```
> for i from 1 to 5 do
```

$$w_{i+1} := \frac{(\mu + \lambda) \cdot w_i - \mu \cdot w_{i-1} - \frac{1}{i}}{\lambda} :$$

```
end do:
```

$$> simplify\left(coeff\left(w_6, \ln\left(\frac{\mu}{\mu - \lambda}\right)\right)\right) \cdot \ln\left(\frac{\mu}{\mu - \lambda}\right)$$

$$+ simplify\left(coeff\left(w_6, \ln\left(\frac{\mu}{\mu - \lambda}\right), 0\right)\right);$$

$$\frac{(\mu + \lambda)\left(\mu^4 + \mu^2\lambda^2 + \lambda^4\right)\ln\left(\frac{\mu}{\mu - \lambda}\right)}{\lambda^6}$$

$$- \frac{1}{60} \frac{60\mu^4 + 90\mu^3\lambda + 110\mu^2\lambda^2 + 125\mu\lambda^3 + 137\lambda^4}{\lambda^5}$$

Can you discern a pattern? (See Example 8.5.)

Similarly, we can set up a difference equation for

$$\tau_i = \mathbb{E}\left(T^2 \mid X(0) = i\right) = \mathbb{E}\left(T_0^2 + 2T_0T_1 + T_1^2 \mid X(0) = i\right),$$

where T_0 is the time till the next transition and T_1 is the *remaining* time till extinction (note T_0 and T_1 are independent), getting

$$\tau_i = \frac{2}{(\lambda + \mu)^2 i^2} + \frac{2}{(\lambda + \mu)i}\left(\frac{\lambda}{\lambda + \mu}\omega_{i+1} + \frac{\mu}{\lambda + \mu}\omega_{i-1}\right)$$

$$+ \frac{\lambda}{\lambda + \mu}\tau_{i+1} + \frac{\mu}{\lambda + \mu}\tau_{i-1}.$$

This enables us to compute τ_{i+1} based on τ_i and τ_{i-1} (and ω_i and ω_{i-1}, which are already known). All we need is $\tau_0 = 0$ and

$$\tau_1 = \int_0^\infty t^2 \cdot \frac{\mathrm{d}r_t}{\mathrm{d}t}\, \mathrm{d}t = w \frac{\mathrm{dilog}\left(\frac{\mu - \lambda}{\mu}\right)}{\lambda\,(\mu - \lambda)},$$

where "dilog" is the dilogarithm function defined in MAPLE as

$$\mathrm{dilog}(x) = \int_1^x \frac{\ln(t)}{1 - t}\, \mathrm{d}t.$$

7.5 LINEAR GROWTH WITH IMMIGRATION

As in the previous section, each member of the process creates an offspring at a rate of λ and perishes at a rate of μ; now we add a stream of immigrants who arrive at an average rate of a.

The corresponding set of difference-differential equations reads

$$\dot{P}_{i,n}(t) = -(\lambda + \mu)n P_{i,n}(t) + \lambda(n - 1)P_{i,n-1}(t) + \mu(n + 1)P_{i,n+1}(t)$$
$$+ a P_{i,n-1}(t) - a P_{i,n}(t) \tag{7.2}$$

correct for all $n \geq 0$ [with the understanding that $P_{i,-1}(t) = 0$]. Multiplying by z^n and summing over n from 0 to ∞ yields

$$\dot{P}_i(z, t) = (\mu - \lambda z)(1 - z)P_i'(z, t) - a(1 - z)P_i(z, t).$$

When $i = 0$, the solution is

$$P_0(z, t) = \left(\frac{p_t}{1 - z q_t}\right)^{a/\lambda},$$

where

$$p_t = \frac{(\lambda - \mu)\mathrm{e}^{-(\lambda - \mu)t}}{\lambda - \mu \mathrm{e}^{-(\lambda - \mu)t}}$$

(the same as before). The resulting distribution is the *modified* (i.e., $X - k$) negative binomial, with parameters $\frac{a}{\lambda}$ (not necessarily an integer) and p_t. This yields the following formula for the individual probabilities

$$\Pr(X(t) = n) = p_t^{a/\lambda}\binom{-\frac{a}{\lambda}}{n}(-q_t)^n$$

and also

$$\mathbb{E}\left(X(t)\right) = \frac{a}{\lambda} \cdot \frac{q_t}{p_t} = a \cdot \frac{1 - e^{-(\lambda-\mu)t}}{(\lambda - \mu)e^{-(\lambda-\mu)t}}$$

$$\text{Var}\left(X(t)\right) = \frac{a}{\lambda} \cdot \frac{q_t}{p_t^2} = a \cdot \frac{\left(1 - e^{-(\lambda-\mu)t}\right) \cdot \left(\lambda - \mu e^{-(\lambda-\mu)t}\right)}{\left((\lambda - \mu)e^{-(\lambda-\mu)t}\right)^2}.$$

At $t = 0$, p_t has the value of 1 (check), at $t = \infty$ we get either $p_t = 0$ (when $\lambda > \mu$) or $p_t = 1 - \frac{\lambda}{\mu}$ (when $\lambda < \mu$). The process thus reaches a *stationary* solution only when $\lambda < \mu$.

When $\lambda = \mu$, p_t reduces to $\frac{1}{1+\mu t}$ ($\Rightarrow q_t = \frac{\mu t}{1+\mu t}$). This implies

$$\mathbb{E}\left(X(t)\right) = at,$$

$$\text{Var}\left(X(t)\right) = at(1 + \mu t)$$

(population explosion at a *linear* rate).

Example 7.2. Take $\lambda = 2.4/\text{h}$, $\mu = 3.8/\text{h}$, $a = 0.9/\text{h}$, and $X(0) = 0$. Find $\mathbb{E}\left(X(1 \text{ day})\right)$, $\text{Var}\left(X(1 \text{ day})\right)$, and $\Pr\left(X(1 \text{ day}) = 3\right)$.

Solution. We quickly discover that $t = 24\,\text{h}$ is, for any practical purposes, large enough to utilize the stationary formulas. We get $q_t = \frac{2.4}{3.8} = 0.63158$, $p_t = \frac{1.4}{3.8} = 0.36842$, and $\frac{a}{\lambda} = 0.375$. The expected value is equal to

$$\frac{a}{\lambda} \cdot \frac{q_t}{p_t} = 0.375 \times \frac{2.4}{1.4} = 0.6429,$$

the corresponding standard deviation is given by

$$\sigma = \sqrt{0.375 \times \frac{2.4}{1.4} \times \left(\frac{3.8}{1.4} - 1\right)} = 1.050,$$

and the probability of having (exactly) three members is

$$\frac{(-0.375) \times (-1.375) \times (-2.375)}{6} \times \left(\frac{1.4}{3.8}\right)^{0.375} \times \left(-\frac{2.4}{3.8}\right)^3 = 3.536\%.$$

$\square$

When $X(0) = i$, we can obtain the complete solution to (7.2) by the following simple argument.

We separate the process into two *independent* processes, the natives and their descendants, and the immigrants with their progeny. The first of these follows the formulas of the previous section, and the second one is the case just studied. Adding them together, we get, for the corresponding PGF,

$$P_i(z,t) = \left(r_t + (1 - r_t)\frac{p_t z}{1 - q_t z}\right)^i \cdot \left(\frac{p_t}{1 - z q_t}\right)^{a/\lambda}.$$

When $\lambda < \mu$, $\lim_{t \to \infty} r_t = 1$. This makes the PGF of the stationary distribution equal to

$$P_i(z, \infty) = \left(\frac{1 - \frac{\lambda}{\mu}}{1 - z\frac{\lambda}{\mu}} \right)^{a/\lambda}$$

and independent of the initial state.

The stationary probabilities, namely,

$$p_n = \Pr(X(\infty) = n),$$

can answer questions about the state of the process in a distant future (in practical terms, the process reaches its stationary state in a handful of time units). They also enable us to compute how frequently any given state is visited in the long run (i.e., when $t \to \infty$).

This is done as follows. After equilibration, the expected value of $I_{X(t)=n}$ (the indicator function of the $X(t) = n$ event, which has a value of one when the process is in State n and a value of zero otherwise) is equal to $\Pr(X(t) = n) \approx p_n$. This means the *empirical* equivalent of $I_{X(t)=n}$ (visualize it, for a specific realization of the process) must have an average value (its integral, divided by the *total* time span) approaching p_n in the long run. This can be rephrased as follows: the long-run *proportion* of time spent in State n must be, to a good approximation, equal to p_n.

Let us now consider the time between two *consecutive* entries to State n (say T_n); this consists of two parts, the time until the process leaves State n (say U_n) followed by the time the process spends among the *other* states before it returns to n. The long-run sum of all such U_n values divided by the sum of all T_n values equals the proportion of time spent in n. Taking the expected value of this ratio, we get

$$\frac{\sum \mathbb{E}(U_n)}{\sum \mathbb{E}(T_n)} \approx \frac{\mathbb{E}(U_n)}{\mathbb{E}(T_n)} \approx p_n.$$

Since we know $\mathbb{E}(U_n) = \frac{1}{\lambda_n + \mu_n}$, we compute

$$\mathbb{E}(T_n) \approx \frac{1}{p_n \cdot (\lambda_n + \mu_n)}.$$

The frequency of visits to State n is the corresponding reciprocal, namely, $p_n \cdot (\lambda_n + \mu_n)$.

Example 7.3. Consider a linear growth with immigration (LGWI) process with *individual* birth and death rates both equal to 0.25/h, an immigration rate of 0.9/h, and an initial value of five natives. Find and plot the distribution of $X(1.35\,\mathrm{h})$.

Solution.

> $e_{\text{aux}} := e^{-t \cdot (\lambda - \mu)}$:

$$> p_t := \begin{cases} \dfrac{1}{1 + \mu \cdot t} & \mu = \lambda \\[3mm] \dfrac{(\lambda - \mu) \cdot e_{\text{aux}}}{\lambda - \mu \cdot e_{\text{aux}}} & \text{otherwise} \end{cases} :$$

$$> r_t := \begin{cases} \dfrac{\mu \cdot t}{1 + \mu \cdot t} & \mu = \lambda \\[3mm] \dfrac{\mu \cdot (1 - e_{\text{aux}})}{\lambda - \mu \cdot e_{\text{aux}}} & \text{otherwise} \end{cases} :$$

$$> P := \left(r_t + \frac{(1 - r_t) \cdot p_t \cdot z}{1 - (1 - p_t) \cdot z} \right)^i \cdot \left(\frac{p_t}{1 - (1 - p_t) \cdot z} \right)^{\frac{a}{\lambda}}$$

$$> (t, \mu, \lambda, a, i) := \left(1.35, \frac{1}{4}, \frac{1}{4}, 1.3, 5 \right) :$$

> $prob := series(P, z, 18)$:

> $pointplot\left([seq\,([j, coeff\,(prob, z, j)], j = 0..17)] \right)$;

$\square$

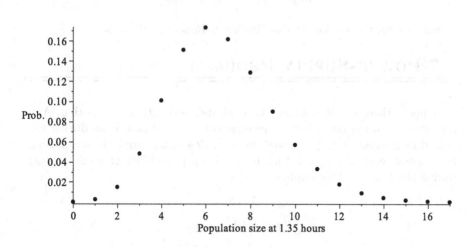

Population size at 1.35 hours

7.6 *M/M/∞* QUEUE

M/M/∞ denotes a queueing system with infinitely many servers, an exponential service time (for each server), and incoming customers forming a Poisson process. An example would be a telephone exchange where phone calls arrive at a constant average rate a and each of them terminates (independently of the rest) with a probability of $\mu\,dt + o(dt)$ during the next interval of length dt (this implies the duration of each phone call is a random variable having an exponential distribution with a mean of $\frac{1}{\mu}$). Clearly, this is a special case of the previous LGWI model, with $\lambda = 0$. Unfortunately, it is not so easy to take the $\lambda \to 0$ limit of the former results; it is easier to simply start from scratch.

Solving

$$\dot{P}_i(z,t) = \mu(1-z)P_i'(z,t) - a(1-z)P_i(z,t)$$

we get (Appendix 7.A)

$$P_i(z,t) = \exp\left(\frac{aq_t}{\mu}\cdot(z-1)\right)\cdot(q_t + p_t z)^i,$$

where $p_t = \mathrm{e}^{-\mu t}$. This corresponds to an *independent sum* of a Poisson-type random variable with a mean of $\frac{aq_t}{\mu}$ and a binomial-type random variable with parameters i and p_t.

We have

$$\mathbb{E}\,(X(t)) = \frac{aq_t}{\mu} + i\cdot p_t$$

and

$$\mathrm{Var}\,(X(t)) = \frac{aq_t}{\mu} + i\cdot p_t q_t.$$

When $t \to \infty$, the stationary distribution is Poisson, with a mean of $\frac{a}{\mu}$.

7.7 POWER-SUPPLY PROBLEM

Suppose there are N welders who work independently of each other. Any one of them, when not using the electric current, will turn it on during the next time interval dt with a probability of $\lambda\cdot dt + o(dt)$. Similarly, when using the current, each welder will turn it off with a probability of $\mu\cdot dt + o(dt)$ during the time dt. This implies

$$\lambda_n = \lambda\cdot(N-n),$$
$$\mu_n = \mu\cdot n,$$

where n represents the number of welders using the current at that moment. We thus get

$$\dot{P}_{i,n}(t) = -\left(\lambda(N-n) + \mu n\right)P_{i,n}(t)$$
$$+ \lambda(N-n+1)P_{i,n-1}(t) + \mu(n+1)P_{i,n+1}(t),$$

implying

$$\dot{P}_i(z,t) = -\lambda N P_i(z,t) - (\mu - \lambda)z P_i'(z,t)$$
$$+ \lambda N z P_i(z,t) - \lambda z^2 P_i'(z,t) + \mu P_i'(z,t)$$
$$= (\mu + \lambda z)(1-z) P_i'(z,t) + N\lambda(z-1) P_i(z,t).$$

The solution is (Appendix 7.A)

$$P_i(z,t) = \left(\frac{\mu - \mu e^{-(\lambda+\mu)t}}{\lambda + \mu} + z \frac{\lambda + \mu e^{-(\lambda+\mu)t}}{\lambda + \mu} \right)^i$$
$$\times \left(\frac{\mu + \lambda e^{-(\lambda+\mu)t}}{\lambda + \mu} + z \frac{\lambda - \lambda e^{-(\lambda+\mu)t}}{\lambda + \mu} \right)^{N-i},$$

which corresponds to an *independent sum* of two random variables, one having a

$$\mathcal{B}\left(i, \frac{\lambda + \mu e^{-(\lambda+\mu)t}}{\lambda + \mu} \right)$$

distribution, the other having a

$$\mathcal{B}\left(N - i, \frac{\lambda - \lambda e^{-(\lambda+\mu)t}}{\lambda + \mu} \right)$$

distribution. As $t \to \infty$, this sum simplifies to $\mathcal{B}\left(N, \frac{\lambda}{\lambda+\mu} \right)$ since $\frac{\lambda}{\lambda+\mu}$ is the limit of both $\frac{\lambda + \mu e^{-(\lambda+\mu)t}}{\lambda+\mu}$ and $\frac{\lambda - \lambda e^{-(\lambda+\mu t)}}{\lambda+\mu}$.

So, at any t, we have

$$\mathbb{E}(X(t)) = i \cdot \frac{\lambda + \mu e^{-(\lambda+\mu)t}}{\lambda + \mu} + (N-i) \cdot \frac{\lambda\left(1 - e^{-(\lambda+\mu)t}\right)}{\lambda + \mu}$$

and

$$\mathrm{Var}\,(X(t)) = i \cdot \frac{\lambda + \mu e^{-(\lambda+\mu)t}}{\lambda + \mu} \cdot \frac{\mu\left(1 - e^{-(\lambda+\mu)t}\right)}{\lambda + \mu}$$
$$+ (N-i) \cdot \frac{\lambda\left(1 - e^{-(\lambda+\mu)t}\right)}{\lambda + \mu} \cdot \frac{\mu + \lambda e^{-(\lambda+\mu)t}}{\lambda + \mu}.$$

7.A SOLVING SIMPLE PDES

Consider

$$\dot{P}(z,t) = a(z) \cdot P'(z,t),$$

where $a(z)$ is a specific (given) function of z.

First we note, if $P(z,t)$ is a solution, then any function of $P(z,t)$, say $g(P(z,t))$, is also a solution. This follows from the chain rule:

$$\dot{g}(P(z,t)) = \acute{g}(P(z,t)) \cdot \dot{P}(z,t),$$
$$g'(P(z,t)) = \acute{g}(P(z,t)) \cdot P'(z,t).$$

Substituted back into the original equation, $\acute{g}(P(z,t))$ (denoting the first derivative of g with respect to its single argument) cancels out.

We will assume

$$P(z,t) = Q(z) \cdot R(t)$$

and substitute this trial solution into the original PDE, getting

$$Q(z) \cdot \dot{R}(t) = a(z) \cdot Q'(z) \cdot R(t).$$

Dividing each side by $Q(z) \cdot R(t)$ results in

$$\frac{\dot{R}(t)}{R(t)} = a(z)\frac{Q'(z)}{Q(z)}.$$

A function of t (but not z) can be equal to a function of z (but not t) only if both OF them are equal to the same constant, say $-\gamma$. We thus get

$$\frac{\dot{R}(t)}{R(t)} = -\gamma$$

and

$$a(z)\frac{Q'(z)}{Q(z)} = -\gamma.$$

The first of these has the following general solution:

$$R(t) = c \cdot e^{-\gamma t};$$

the second one implies

$$\ln Q(z) = -\gamma \int \frac{dz}{a(z)}$$

or

$$Q(z) = \exp\left(-\gamma \int \frac{dz}{a(z)}\right).$$

We then know

$$g\left(c \cdot e^{-\gamma t} \cdot \exp\left(-\gamma \int \frac{dz}{a(z)}\right)\right)$$

is also a solution, where g is any univariate function. Clearly, both the multiplication by c and raising to the power of γ can be absorbed into g, so rewriting the solution as

$$g\left(e^{-t} \cdot \exp\left(-\int \frac{dz}{a(z)}\right)\right)$$

is an equivalent way of putting it. Furthermore, one can show this represents the *general* solution of the original equation (i.e., all of its solutions have this form).

The initial condition

$$P(z,0) = z^i = g\left(\exp\left(-\int \frac{dz}{a(z)}\right)\right)$$

(where i is the value of the process at time $t = 0$) then determines the specific form of $g(\cdots)$.

Example 7.4. Yule process: $a(z) = -\lambda z(1 - z)$. Since

$$\int \frac{dz}{z(1-z)} = \int \left(\frac{1}{z} + \frac{1}{1-z}\right) dz = \ln z - \ln(1 - z),$$

we get

$$P(z,t) = g_0\left(e^{-t} \cdot \exp\left(\frac{1}{\lambda} \ln \frac{z}{1-z}\right)\right)$$

$$= g_0\left(e^{-t} \cdot \left(\frac{z}{1-z}\right)^{1/\lambda}\right)$$

$$= g\left(e^{-\lambda t} \cdot \frac{z}{1-z}\right),$$

where $g(\cdots)$ is such that

$$g\left(\frac{z}{1-z}\right) = z^i$$

or

$$g(x) = \left(\frac{x}{1+x}\right)^i.$$

The final solution is thus

$$P(z,t) = \left(\frac{e^{-\lambda t} \cdot \frac{z}{1-z}}{1 + e^{-\lambda t} \cdot \frac{z}{1-z}} \right)^i = \left(\frac{e^{-\lambda t} \cdot z}{1 - z + e^{-\lambda t} \cdot z} \right)^i .$$

This is the PGF of a negative binomial distribution (number of trials to achieve the ith success) with $p_t = e^{-\lambda t}$. ◇

Example 7.5. Pure-death process: $a(z) = \mu(1 - z)$.

$$P(z,t) = g_0 \left(e^{-t} \cdot \exp\left(\frac{1}{\mu} \ln(1 - z) \right) \right)$$

$$= g_0 \left(e^{-t} \cdot (1 - z)^{1/\mu} \right)$$

$$= g \left(e^{-\mu t} \cdot (1 - z) \right),$$

where $g(\cdots)$ is such that

$$g(1 - z) = z^i$$

or

$$g(x) = (1 - x)^i .$$

The final solution is thus

$$P(z,t) = \left(1 - e^{-\mu t} \cdot (1 - z) \right)^i = (1 - e^{-\mu t} + e^{-\mu t} \cdot z)^i .$$

This is the PGF of a binomial distribution, with $p_t = e^{-\mu t}$ the and total number of trials equal to i. ◇

Example 7.6. Linear-growth process: $a(z) = (1 - z)(\mu - \lambda z)$. Since

$$\int \frac{dz}{(1 - z)(\mu - \lambda z)} = \frac{1}{\lambda - \mu} \int \left(\frac{\lambda}{\mu - \lambda z} - \frac{1}{1 - z} \right) dz$$

$$= \frac{1}{\lambda - \mu} (\ln(1 - z) - \ln(\mu - \lambda z)),$$

we get

$$P(z,t) = g_0 \left(e^{-t} \cdot \left(\frac{\mu - \lambda z}{1 - z} \right)^{\frac{1}{\lambda - \mu}} \right) = g \left(e^{-t(\lambda - \mu)} \cdot \frac{\mu - \lambda z}{1 - z} \right),$$

where $g(\cdots)$ is such that

$$g \left(\frac{\mu - \lambda z}{1 - z} \right) = z^i$$

or

$$g(x) = \left(\frac{\mu - x}{\lambda - x} \right)^i.$$

The final solution is thus

$$P(z,t) = \left(\frac{\mu - e^{-t(\lambda - \mu)} \cdot \frac{\mu - \lambda z}{1 - z}}{\lambda - e^{-t(\lambda - \mu)} \cdot \frac{\mu - \lambda z}{1 - z}} \right)^i$$

$$= \left(\frac{\mu(1 - z) - e^{-t(\lambda - \mu)} \cdot (\mu - \lambda z)}{\lambda(1 - z) - e^{-t(\lambda - \mu)} \cdot (\mu - \lambda z)} \right)^i$$

$$= \left(\frac{\mu \left(1 - e^{-t(\lambda - \mu)} \right) - \left(\mu - \lambda e^{-t(\lambda - \mu)} \right) z}{\lambda - \mu e^{-t(\lambda - \mu)} - \lambda \left(1 - e^{-t(\lambda - \mu)} \right) z} \right)^i$$

$$= \left(\frac{\frac{\mu \left(1 - e^{-t(\lambda - \mu)} \right)}{\lambda - \mu e^{-t(\lambda - \mu)}} - \frac{\mu - \lambda e^{-t(\lambda - \mu)}}{\lambda - \mu e^{-t(\lambda - \mu)}}}{1 - \frac{\lambda \left(1 - e^{-t(\lambda - \mu)} \right)}{\lambda - \mu e^{-t(\lambda - \mu)}} z} \right)^i$$

$$= \left(\frac{r_t - (r_t + q_t - 1)z}{1 - q_t z} \right)^i$$

$$= \left(\frac{r_t - r_t(q_t + p_t)z + p_t z}{1 - q_t z} \right)^i$$

$$= \left(r_t + (1 - r_t) \frac{p_t z}{1 - q_t z} \right)^i.$$

This is a composition of a binomial distribution with i trials and a success probability of $1 - r_t$ and of a geometric distribution with a probability of success equal to p_i. $\diamond$

EXTENSION

We now solve a slightly more complicated PDE, namely,

$$\dot{P}(z,t) = a(z) \cdot P'(z,t) + b(z) \cdot P(z,t).$$

One can show the general solution can be found by solving the homogeneous version of this equation first [without the last term; we already know how to do this – let us denote it by $G(z,t)$] and then multiplying it by a function of z, say $h(z)$. Substituting this trial solution into the full (nonhomogeneous) equation one can solve for $h(z)$:

$$\dot{G}(z,t)h(z) = a(z)G'(z,t)h(z) + a(z)G(z,t)h'(z) + b(z)G(z,t), h(z).$$

Since

$$\dot{G}(z,t) = a(z)G'(z,t)$$

(by our original assumption), we can cancel the first two terms and write

$$0 = a(z)G(z,t)h'(z) + b(z)G(z,t), h(z),$$

implying

$$\frac{h'(z)}{h(z)} = -\frac{b(z)}{a(z)},$$

which can be easily solved:

$$h(z) = \exp\left(-\int \frac{b(z)}{a(z)}\,dz\right).$$

To meet the initial-value condition, we must find $g(\cdots)$ such that

$$g\left(\exp\left(-\int \frac{dz}{a(z)}\right)\right) \cdot h(z) = z^i.$$

Example 7.7. Linear growth with immigration:

$$a(z) = (1-z)(\mu - \lambda z),$$
$$b(z) = -a(1-z).$$

We have already solved the homogeneous version, so let us find

$$h(z) = \exp\left(a \int \frac{1-z}{(1-z)(\mu - \lambda z)}\,dz\right)$$
$$= \exp\left(-\frac{a}{\lambda}\ln(\mu - \lambda z)\right)$$
$$= (\mu - \lambda z)^{-a/\lambda}.$$

The general solution is thus

$$g\left(e^{-t(\lambda-\mu)} \cdot \frac{\mu - \lambda z}{1-z}\right)(\mu - \lambda z)^{-a/\lambda}.$$

When $i = 0$,

$$g\left(\frac{\mu - \lambda z}{1-z}\right)(\mu - \lambda z)^{-a/\lambda} = 1,$$

$$g\left(\frac{\mu - \lambda z}{1-z}\right) = (\mu - \lambda z)^{a/\lambda},$$

$$g(x) = \left(\mu - \lambda\frac{\mu - x}{\lambda - x}\right)^{a/\lambda} = \left(\frac{(\lambda - \mu)x}{\lambda - x}\right)^{a/\lambda},$$

resulting in

$$\left(\frac{(\lambda - \mu)e^{-t(\lambda-\mu)} \cdot \frac{\mu-\lambda z}{1-z}}{\lambda - e^{-t(\lambda-\mu)} \cdot \frac{\mu-\lambda z}{1-z}} \right)^{a/\lambda} (\mu - \lambda z)^{-a/\lambda}$$

$$= \left(\frac{(\lambda - \mu)e^{-t(\lambda-\mu)}}{\lambda(1-z) - e^{-t(\lambda-\mu)}(\mu - \lambda z)} \right)^{a/\lambda}$$

$$= \left(\frac{(\lambda - \mu)e^{-t(\lambda-\mu)}}{\lambda - \mu e^{-t(\lambda-\mu)} - \lambda(1 - e^{-t(\lambda-\mu)})z} \right)^{a/\lambda}$$

$$= \left(\frac{\frac{(\lambda - \mu)e^{-t(\lambda-\mu)}}{\lambda - \mu e^{-t(\lambda-\mu)}}}{1 - \frac{\lambda(1 - e^{-t(\lambda-\mu)})}{\lambda - \mu e^{-t(\lambda-\mu)}} z} \right)^{a/\lambda}$$

$$= \left(\frac{p_t}{1 - q_t z} \right)^{a/\lambda},$$

which is the *modified* negative binomial distribution with parameters p_t and $\frac{a}{\lambda}$. $\diamondsuit$

Example 7.8. M/M/∞ queue:

$$a(z) = \mu(1 - z),$$
$$b(z) = -a(1 - z).$$

The general solution to the homogeneous version of the equation is

$$g_0 \left(e^{-t} \cdot \exp\left(-\frac{1}{\mu} \int \frac{dz}{1-z} \right) \right) = g_0 \left(e^{-t} \cdot (1-z)^{1/\mu} \right)$$
$$= g \left(e^{-\mu t} \cdot (1-z) \right).$$

Then we get

$$h(z) = \exp\left(\frac{a}{\mu} \int dz \right) = \exp\left(\frac{a}{\mu} z \right).$$

The general solution is thus

$$g \left(e^{-\mu t} \cdot (1-z) \right) \cdot \exp\left(\frac{a}{\mu} z \right).$$

To meet the usual initial condition, we need

$$g(1 - z) \cdot \exp\left(\frac{a}{\mu} z \right) = z^i$$

or

$$g(1-z) = \exp\left(-\frac{a}{\mu}z\right) \cdot z^i,$$

implying

$$g(x) = \exp\left(-\frac{a}{\mu}(1-x)\right) \cdot (1-x)^i.$$

Finally, we must replace x by the original argument of g and multiply by $h(z)$:

$$\exp\left(-\frac{a}{\mu}\left(1 - e^{-\mu t} \cdot (1-z)\right)\right) \cdot \left(1 - e^{-\mu t} \cdot (1-z)\right)^i \cdot \exp\left(\frac{a}{\mu}z\right)$$

$$= \exp\left(-\frac{a}{\mu}\left(1 - e^{-\mu t} + e^{-\mu t}z - z\right)\right) \cdot (q_t + p_t z)^i$$

$$= \exp\left(-\frac{a \cdot q_t}{\mu}(1-z)\right) \cdot (q_t + p_t z)^i,$$

where $p_t = e^{-\mu t}$. This is an independent sum (a convolution) of a Poisson distribution having a mean of $\Lambda = \frac{a \cdot q_t}{\mu}$ and a binomial distribution with parameters i and p_t. $\diamond$

Example 7.9. Power-supply process (N welders):

$$a(z) = (1-z)(\mu + \lambda z),$$
$$b(z) = -N\lambda(1-z).$$

Since

$$\int \frac{dz}{(1-z)(\mu + \lambda z)} = \frac{1}{\mu + \lambda}\int\left(\frac{1}{1-z} + \frac{\lambda}{\mu + \lambda z}\right) dz$$

$$= \frac{1}{\lambda + \mu}\left(\ln(\mu + \lambda z) - \ln(1-z)\right),$$

the general solution to the homogeneous version of the equation is

$$g_0\left(e^{-t} \cdot \left(\frac{1-z}{\mu + \lambda z}\right)^{\frac{1}{\lambda + \mu}}\right) = g\left(e^{-t(\lambda+\mu)} \cdot \frac{1-z}{\mu + \lambda z}\right).$$

Then we get

$$h(z) = \exp\left(N\lambda \int \frac{dz}{\mu + \lambda z}\right) = (\mu + \lambda z)^N.$$

The general solution is thus

$$g\left(e^{-t(\lambda+\mu)} \cdot \frac{1-z}{\mu+\lambda z}\right) \cdot (\mu + \lambda z)^N.$$

To meet the initial condition, we need

$$g\left(\frac{1-z}{\mu+\lambda z}\right) \cdot (\mu + \lambda z)^N = z^i$$

or

$$g\left(\frac{1-z}{\mu+\lambda z}\right) = z^i \cdot (\mu + \lambda z)^{-N},$$

implying

$$g(x) = \left(\frac{1-\mu x}{1+\lambda x}\right)^i \cdot \left(\frac{\lambda+\mu}{1+\lambda x}\right)^{-N}.$$

Finally, we must replace x by the original argument of g and multiply by $h(z)$:

$$\left(\frac{1 - \mu e^{-t(\lambda+\mu)}\frac{1-z}{\mu+\lambda z}}{1 + \lambda e^{-t(\lambda+\mu)}\frac{1-z}{\mu+\lambda z}}\right)^i \left(\frac{1 + \lambda e^{-t(\lambda+\mu)}\frac{1-z}{\mu+\lambda z}}{\lambda+\mu}\right)^N (\mu + \lambda z)^N$$

$$= \left(\frac{\mu + \lambda z - \mu e^{-t(\lambda+\mu)}(1-z)}{\mu + \lambda z + \lambda e^{-t(\lambda+\mu)}(1-z)}\right)^i \times \left(\frac{\mu + \lambda z + \lambda e^{-t(\lambda+\mu)}(1-z)}{\lambda+\mu}\right)^N$$

$$= \left(\frac{\mu(1 - e^{-t(\mu+\lambda)}) + (\lambda + \mu e^{-t(\mu+\lambda)})z}{\lambda+\mu}\right)^i$$

$$\times \left(\frac{\mu + \lambda e^{-t(\mu+\lambda)} + \lambda(1 - e^{-t(\mu+\lambda)})z}{\lambda+\mu}\right)^{N-i}$$

$$= \left(q_t^{(1)} + p_t^{(1)}z\right)^i \times \left(q_t^{(2)} + p_t^{(2)}z\right)^{N-i}$$

(a convolution of two binomials), where

$$p_t^{(1)} = \frac{\lambda + \mu e^{-t(\mu+\lambda)}}{\lambda+\mu},$$

$$p_t^{(2)} = \frac{\lambda(1 - e^{-t(\mu+\lambda)})}{\lambda+\mu}.$$

Note the same answer could have been obtained by taking the complete ($i \neq 0$) solution of Example 7.7 and replacing λ by $-\lambda$ and a by $N - \lambda$. ◇

EXERCISES

Exercise 7.1. Consider a pure-birth Markov process with $\lambda_n = 2.34 \times n$ per hour and an initial value of $X(0) = 4$. Find:

(a) $\Pr\left(X(36\,\text{min}) \leq 10\right)$;
(b) $\mathbb{E}\left(X(16\,\text{min}\,37\,\text{s})\right)$ and the corresponding standard deviation.

Exercise 7.2. Consider a pure-death Markov process with $\mu_n = 2.34 \times n$ per hour and an initial value of $X(0) = 33$. Find:

(a) $\Pr\left(X(36\,\text{min}) \leq 10\right)$;
(b) $\mathbb{E}\left(X(16\,\text{min}\,37\,\text{s})\right)$ and the corresponding standard deviation;
(c) The probability that the process will become extinct *during* its second hour;
(d) The expected time until extinction and the corresponding standard deviation.

Exercise 7.3. Consider a linear-growth process with the following rates:

$$\lambda_n = 3n \text{ per hour,}$$
$$\mu_n = 4n \text{ per hour,}$$

and an initial value of three members. Find:

(a) The probability that 30 min later the process will have more than four members;
(b) The corresponding (i.e., 30 min later) mean and standard deviation of the value of the process;
(c) The probability that the process will become extinct during the first 20 min;
(d) The expected time until extinction and the corresponding standard deviation.

Exercise 7.4. Consider a linear-growth process with the following rates:

$$\lambda_n = 4n \text{ per hour,}$$
$$\mu_n = 3n \text{ per hour,}$$

and an initial value of three members. Find:

(a) The probability that 30 min later the process will have more than four members;
(b) The corresponding (i.e., 30 min later) expected value and standard deviation;

(c) The probability that the process will become extinct during the first 20 min;

(d) The probability of ultimate extinction.

Exercise 7.5. Consider the following PDE:

$$z \, \dot{P}(z,t) = (z-1) P'(z,t).$$

(a) Find its general solution (assume $z < 1$).

(b) Find the solution that satisfies the condition $P(z,0) = (1-z)e^z$.

Exercise 7.6. Consider a birth-and-death (B&D) process with the following rates:

$$\lambda_n = (27 - 3n) \text{ per hour},$$
$$\mu_n = 5n \text{ per hour},$$

where $n = 0, 1, 2, \ldots, 9$.

(a) If the process starts in State 4, what is the probability that 8 h later the process will be in State 5?

(b) What is the expected time between two consecutive visits to State 0 (entry to entry)?

Exercise 7.7. Consider a LGWI process with the following rates:

$$\lambda_n = (8.12\,n + 2.43) \text{ per hour},$$
$$\mu_n = (9.04\,n) \text{ per hour},$$

and consisting of 13 members at 8:27. Compute:

(a) The expected value of the process at 9:42 a.m. and the corresponding standard deviation;

(b) The probability that at 9:42 a.m. the process will have more than 15 members;

(c) The expected time until the extinction of the native subpopulation (initial members and their descendants) and the corresponding standard deviation;

(d) The long-run frequency of visits to State 0 (per day) and their average duration (in minutes and seconds).

Exercise 7.8. Consider a B&D process with the following rates:

$$\lambda_n = (72 - 8n) \text{ per hour},$$
$$\mu_n = 11n \text{ per hour},$$

and the value of 7 at $t = 0$. Compute:

(a) The probability that all of the next three transitions will be deaths;
(b) The expected value of the process at $t = 6\,\mathrm{min}$ and the corresponding standard deviation;
(c) The probability that at $t = 6\,\mathrm{min}$ the process will have a value smaller than 4;
(d) The long-run frequency of visits to State 0 (per day) and their average duration (in seconds).

Exercise 7.9. Consider an $M/M/\infty$ queue with customers arriving at a rate of 25.3 per hour, an expected service time of 17 min, and 9 customers being serviced at 10:17 a.m. Compute:

(a) The probability that all of these 9 customers will have finished their service by 11:00 a.m. (note we are ignoring new arrivals);
(b) The expected number of customers (including new arrivals) being serviced at 10:25 a.m. and the corresponding standard deviation;
(c) The probability that at 10:25 a.m. fewer than eight customers will be getting serviced;
(d) The long-run frequency of visits to State 0 (per week) and their average duration (in minutes and seconds).

Exercise 7.10. Find the general solution to

$$\dot{P}(z,t) = P'(z,t) + P(z,t)$$

(assume $0 \le z < 1$). Also, find the specific solution that satisfies

$$P(z,0) = z.$$

CHAPTER 8
Birth-and-Death Processes II

We investigate birth-and-death (B&D) processes tht are too complicated (in terms of the λ_n and μ_n rates) to have a full analytic solution. We settle for studying only their LONG-RUN behavior. Specifically, we are interested in finding either the corresponding stationary distribution or, when state 0 is absorbing (e.g., a population can go extinct), the probability of ultimate extinction.

8.1 CONSTRUCTING A STATIONARY DISTRIBUTION

So far we have dealt with models having a full, analytic solution in terms of $P_{i,n}(t)$. This was made possible because both λ_n and μ_n were rather simple linear functions of n. For more complicated models, an explicit analytic solution may be impossible to find.

What we can still do in that case (when there are no absorbing states) is to construct a stationary solution (which covers most of the process's future behavior). This can be done by assuming $t \to \infty$, which implies $P_{i,n}(t) \to p_n$ (independent of the initial state). Stationary probabilities must meet the following set of equations:

$$\lambda_0 p_0 = \mu_1 p_1$$
$$(\lambda_1 + \mu_1)p_1 = \lambda_0 p_0 + \mu_2 p_2,$$
$$(\lambda_2 + \mu_2)p_2 = \lambda_1 p_1 + \mu_3 p_3,$$
$$(\lambda_3 + \mu_3)p_3 = \lambda_2 p_2 + \mu_4 p_4$$
$$\vdots$$

J. Vrbik and P. Vrbik, *Informal Introduction to Stochastic Processes with Maple*, Universitext, DOI 10.1007/978-1-4614-4057-4_8,
© Springer Science+Business Media, LLC 2013

(a set of difference equations, but this time the coefficients are not constant). Assuming p_0 is known, we can solve these recursively as follows:

$$p_1 = \frac{\lambda_0}{\mu_1} p_0,$$

$$p_2 = \frac{1}{\mu_2}\left(\frac{\lambda_1 + \mu_1}{\mu_1}\lambda_0 p_0 - \lambda_0 p_0\right) = \frac{\lambda_0\lambda_1}{\mu_1\mu_2}p_0,$$

$$p_3 = \frac{1}{\mu_3}\left(\frac{\lambda_2 + \mu_2}{\mu_2}\cdot\frac{\lambda_1\lambda_0}{\mu_1}p_0 - \frac{\lambda_1\lambda_0}{\mu_1}p_0\right) = \frac{\lambda_0\lambda_1\lambda_2}{\mu_1\mu_2\mu_3}p_0,$$

$$p_4 = \frac{\lambda_0\lambda_1\lambda_2\lambda_3}{\mu_1\mu_2\mu_3\mu_4}p_0,$$

$$\vdots$$

$$p_n = \frac{\lambda_0\lambda_1\lambda_2\ldots\lambda_{n-1}}{\mu_1\mu_2\mu_3\ldots\mu_n}p_0.$$

We know the sum of these is 1, and thus

$$p_0 = \left(1 + \sum_{n=1}^{\infty}\frac{\lambda_0\lambda_1\lambda_2\cdots\lambda_{n-1}}{\mu_1\mu_2\mu_3\cdots\mu_n}\right)^{-1}.$$

When the sum *diverges*, a stationary distribution does *not* exist and the process is bound for population explosion.

Thus, we have a relatively simple procedure for constructing a stationary distribution. Note the aforementioned sum cannot diverge when there are only finitely many states.

Example 8.1. In the case of N welders, the procedure would work as follows:

n	0	1	2	3	$\cdots$	$N-1$	N
λ_n	$N\lambda$	$(N-1)\lambda$	$(N-2)\lambda$	$(N-3)\lambda$	$\cdots$	λ	0
μ_n	0	μ	2μ	3μ	$\cdots$	$(N-1)\mu$	$N\mu$
$\frac{\lambda_{n-1}}{\mu_n}$	$\times$	$N\rho$	$\frac{N-1}{2}\rho$	$\frac{N-2}{3}\rho$	$\cdots$	$\frac{2}{N-1}\rho$	$\frac{1}{N}\rho$
$\frac{p_n}{p_0}$	1	$N\rho$	$\binom{N}{2}\rho^2$	$\binom{N}{3}\rho^3$	$\cdots$	$\binom{N}{N-1}\rho^{N-1}$	$\binom{N}{N}\rho^N$

where $\rho = \frac{\lambda}{\mu}$. Because the sum of the quantities in the last row is $(1+\rho)^N$, the answer is

$$p_n = \binom{N}{n}\frac{\rho^n}{(1+\rho)^N} = \binom{N}{n}\left(\frac{\rho}{1+\rho}\right)^n\left(\frac{1}{1+\rho}\right)^{N-n},$$

where $0 \le n \le N$, which agrees with our previous result. $\diamond$

The expected value of this distribution is $N \cdot \frac{\lambda}{\lambda + \mu}$, which represents

1. The expected number of welders using electricity when t is large and
2. The time-averaged number of welders using electricity in the *long run*.

Example 8.2. Find and display the stationary distribution (verify it does exist) of a B&D process with

$$\lambda_n = 15 \times 0.8^n,$$

$$\mu_n = \frac{6n}{1 + 0.3n}.$$

Solution.

> $\lambda := n \rightarrow 15 \cdot 0.8^n$:

> $\mu := n \rightarrow \dfrac{6 \cdot n}{1 + 0.3 \cdot n}$:

> $S := evalf\left(\displaystyle\sum_{m=0}^{\infty} \prod_{k=0}^{m-1} \dfrac{\lambda(k)}{\mu(k+1)} \right)$;

$$S := 21.3214$$

> $pointplot\left(\left[seq\left(\left[m, \dfrac{1}{S} \cdot \displaystyle\prod_{k=0}^{m-1} \dfrac{\lambda(j)}{\mu(k+1)} \right], m = 0..9 \right) \right] \right)$;

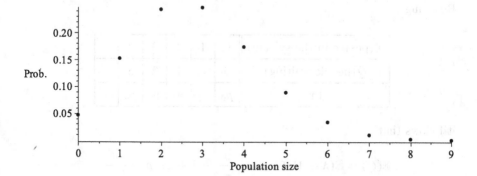

MORE EXAMPLES

$M/M/1$ QUEUE

This time we have only one server, which means most customers will have to line up and wait for service. Here, $\lambda_n = \lambda$ for all $n \geq 0$, $\mu_n = \mu$ for $n \geq 1$, and $\mu_0 = 0$.

The following table will help us build the stationary probabilities:

State	0	1	2	3	$\cdots$
λ_n	λ	λ	λ	λ	$\cdots$
μ_n	0	μ	μ	μ	$\cdots$
$\frac{\lambda_{n-1}}{\mu_n}$	$\times$	ρ	ρ	ρ	$\cdots$
$\frac{p_n}{p_0}$	1	ρ	ρ^2	ρ^3	$\cdots$

where $\rho = \frac{\lambda}{\mu}$. Thus, $p_n = \rho^n(1 - \rho)$ when $\lambda < \mu$ (the queue is unstable; otherwise it keeps on growing). One can recognize this as the modified geometric distribution with its basic parameter (we used to call it p) equal to $1 - \rho$. The average number of people in the system is thus

$$\frac{q}{p} = \frac{\rho}{1 - \rho} = \frac{\lambda}{\mu - \lambda}.$$

What is the average number of people waiting (in the actual queue)? Realizing

X (people in the system)	0	1	2	3	4	$\cdots$
Q (people waiting)	0	0	1	2	3	$\cdots$
Pr	p_0	p_1	p_2	p_3	p_4	$\cdots$

it follows that

$$\mathbb{E}(Q) = \mathbb{E}(X - 1) + p_0 = \frac{\rho}{1 - \rho} - 1 + 1 - \rho = \frac{\rho^2}{1 - \rho}.$$

The proportion of time the server is busy (SERVER UTILIZATION FACTOR) is equal to $1 - p_0 = \rho$; the average length of an IDLE PERIOD is $\frac{1}{\lambda}$.

The average length of a BUSY CYCLE (from one idle period to the next, measured from end to end) is thus

$$\frac{\frac{1}{\lambda}}{p_0} = \frac{\mu}{\lambda(\mu - \lambda)}.$$

(the average length of an idle period divided by the percentage of time the server is idle); the average length of a BUSY PERIOD (the same thing, measured from end to the beginning of two consecutive idle periods) is thus

$$\frac{\mu}{\lambda(\mu - \lambda)} - \frac{1}{\lambda} = \frac{1}{\mu - \lambda}.$$

$M/M/1$ QUEUE WITH BALKING

Now we introduce α_n, the probability of a new customer staying given he finds n people in the system (with a probability of $1 - \alpha_n$, he leaves or *balks*). Here we must modify only the λ_n rates; thus, $\lambda_n = \lambda \cdot \alpha_n$. Note the probability that a new virtual arrival (regardless of whether this customer stays or leaves) will find n customers in the system (waiting or being served) is given by the corresponding *stationary* probability p_n.

There are many special cases of this situation, one of which is that of a FINITE WAITING ROOM, where $\alpha_n = 1$ when $n \leq N$ and $\alpha_n = 0$ otherwise (the waiting room can accommodate only $N - 1$ people; people who can fit stay, those who find it full must leave and not return).

We will leave questions of this type for the exercises.

$M/M/c$ QUEUE

With c servers, all the λ_n are again equal to a constant λ, but $\mu_n = n \cdot \mu$ for $n \leq c$ and $\mu_n = c \cdot \mu$ when $n \geq c$ (all servers are busy). Thus, we have

State	0	1	2	3	$\cdots$	c	$c+1$	$c+2$	$\cdots$
λ_n	λ	λ	λ	λ	$\cdots$	λ	λ	λ	$\cdots$
μ_n	0	μ	2μ	3μ	$\cdots$	$c\mu$	$c\mu$	$c\mu$	$\cdots$
$\frac{\lambda_{n-1}}{\mu_n}$	$\times$	ρ	$\frac{\rho}{2}$	$\frac{\rho}{3}$	$\cdots$	$\frac{\rho}{c}$	$\frac{\rho}{c}$	$\frac{\rho}{c}$	$\cdots$
$\frac{p_n}{p_0}$	1	ρ	$\frac{\rho^2}{2!}$	$\frac{\rho^3}{3!}$	$\cdots$	$\frac{\rho^c}{c!}$	$\frac{\rho^c}{c!}\frac{\rho}{c}$	$\frac{\rho^c}{c!}(\frac{\rho}{c})^2$	$\cdots$

Now,

$$p_n = \begin{cases} \dfrac{\rho^n}{n!\Gamma} & n \leq c, \\[2ex] \dfrac{\rho^n}{c!\Gamma c^{n-c}} & n \geq c, \end{cases}$$

where $\rho = \frac{\lambda}{\mu}$ and

$$\Gamma = \sum_{k=0}^{c-1} \frac{\rho^k}{k!} + \sum_{k=c}^{\infty} \frac{\rho^k}{c!c^{k-c}} = \sum_{k=0}^{c-1} \frac{\rho^k}{k!} + \frac{\rho^c}{c!(1 - \frac{\rho}{c})},$$

provided $\lambda < c\mu$ (otherwise, the queue is unstable).

The average number of BUSY SERVERS is (visualize the corresponding table)

$$\frac{\sum\limits_{k=1}^{c} \dfrac{\rho^k}{(k-1)!} + \dfrac{\rho^{c+1}}{c!(1-\frac{\rho}{c})}}{\Gamma} = \rho.$$

This, divided by c, yields the average proportion of busy servers defining the corresponding SERVER UTILIZATION FACTOR; in this case it is equal to $\frac{\rho}{c}$.

Similarly, the average size of the actual queue (again, visualize the corresponding table) is

$$\frac{\rho^c}{c!\Gamma} \sum_{i=1}^{\infty} i \left(\frac{\rho}{c}\right)^i = \frac{\rho^c}{c!\Gamma} \cdot \frac{\frac{\rho}{c}}{(1-\frac{\rho}{c})^2}.$$

8.2 LITTLE'S FORMULAS

One can show that, in general, for *any* queuing system we must have

$$\mathbb{E}(X_\infty) = \lambda_{\text{av}} \cdot \mathbb{E}(U),$$

where X_∞ is the number of customers and U is the total time spent in the system (waiting *and* being serviced) by a customer, after the process has reached its *stationary* state. The two expected values can then be interpreted as long-run averages; similarly, λ_{av} is the long-run average rate of arrivals.

The correctness of this formula can be demonstrated by comparing the following two graphs: the first displays the current value of X_t for a time period of length T (we use $T = 1$, but one must visualize extensions of these as T increases), and the second one shows both the arrival and departure time of each customer (the beginning and end of each box; the boxes are of height 1 and move up one unit with each new customer).

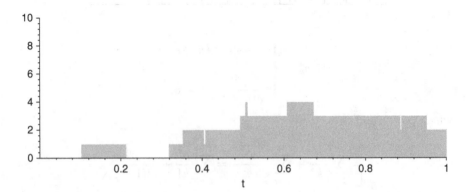

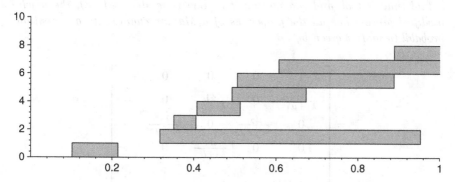

The total height of all boxes at time t yields a value of X_t, implying both graphs must have the same shaded area.

The area of the first graph, divided by T, tends, in the $T \to \infty$ limit, to $\mathbb{E}(X_\infty)$, that is, the long-run average of X_t.

The second graph has, to a good approximation, $\lambda_{\mathrm{av}} \cdot T$ boxes (true in the $T \to \infty$ limit) of the average length $\mathbb{E}(U)$. The total area, divided by T, must thus be equal (in the same limit) to $\lambda_{\mathrm{av}} \cdot \mathbb{E}(U)$.

We can modify the two graphs (or, rather, their interpretation) by replacing X_t by Q_t (number of waiting customers) and U by W (a customer's waiting time). Note in this case, some of the boxes may have zero length (a lucky customer does not have to wait at all). Using the same kind of argument, we can prove

$$\mathbb{E}(Q_\infty) = \lambda_{\mathrm{av}} \cdot \mathbb{E}(W).$$

Finally, by subtracting the two equations, we get

$$\mathbb{E}(Y_\infty) = \lambda_{\mathrm{av}} \cdot \mathbb{E}(S),$$

where Y_∞ is the number of customers being serviced (i.e., the number of *busy* servers) and S is a service time.

With the help of these formulas, one can bypass some of the tedious proofs of the previous section.

8.3 ABSORPTION ISSUES

When State 0 is absorbing, the stationary distribution is degenerate (concentrated at 0), and the only nontrivial issues are to find the probability of absorption (i.e., extinction, in this context) and the expected time till it occurs (when certain). To deal with these problems, we first introduce a new concept of an EMBEDDED MARKOV CHAIN (EMC).

Proposition 8.1. *Consider a B&D process with an absorbing State 0. When each jump from one state to another is seen as a transition (ignoring the* *actual time it took and considering it as one time step instead), the newly modified process has all the properties of a Markov chain, with a transition probability matrix given by*

$$
\mathbb{P} =
\begin{bmatrix}
1 & 0 & 0 & 0 & \cdots \\
\frac{\mu_1}{\lambda_1+\mu_1} & 0 & \frac{\lambda_1}{\lambda_1+\mu_1} & 0 & \cdots \\
0 & \frac{\mu_2}{\lambda_2+\mu_2} & 0 & \frac{\lambda_2}{\lambda_2+\mu_2} & \cdots \\
0 & 0 & \frac{\mu_3}{\lambda_3+\mu_3} & 0 & \cdots \\
\vdots & \vdots & \vdots & \vdots & \ddots
\end{bmatrix}.
$$

Proof. Suppose a B&D process is in State i. Define X (Y) as the time till the next jump up (down). From what we know already, each is a random variable with a mean of $\frac{1}{\lambda_i}$ ($\frac{1}{\mu_i}$). The process actually takes a transition corresponding to the smaller one of these two (the other value is discarded), and the same competition starts all over again. Clearly, $Z \equiv \min(X,Y)$ is the time till the next jump; the probability it will have a value greater than z is $\Pr(X > z \cap Y > z) = e^{-\lambda_i z} \cdot e^{-\mu_i z} = e^{-(\lambda_i+\mu_i)z}$, and Z is thus exponential with a mean of $\frac{1}{\lambda_i+\mu_i}$.

Furthermore,

$$
\Pr(X > Y | Z = z)
$$
$$
= \lim_{\Delta \to 0} \Pr(X > Y | z \le Z < z + \Delta)
$$
$$
= \frac{\lambda_i \mu_i \int\limits_{z}^{z+\Delta} \int\limits_{y}^{\infty} e^{-\lambda_i x - \mu_i y} \, dx \, dy}{\lambda_i \mu_i \int\limits_{z}^{z+\Delta} \int\limits_{y}^{\infty} e^{-\lambda_i x - \mu_i y} dx dy + \lambda_i \mu_i \int\limits_{z}^{z+\Delta} \int\limits_{x}^{\infty} e^{-\lambda_i x - \mu_i y} \, dy \, dx}
$$
$$
= \frac{\mu_i \int\limits_{z}^{z+\Delta} e^{-\lambda_i y - \mu_i y} \, dy}{\mu_i \int\limits_{z}^{z+\Delta} e^{-\lambda_i y - \mu_i y} \, dy + \lambda_i \int\limits_{z}^{z+\Delta} e^{-\lambda_i x - \mu_i x} \, dx}
$$
$$
= \frac{\mu_i}{\mu_i + \lambda_i}
$$

regardless of the value of z. □

Note, based on this EMC, we can recover the original Markov process if we are given the values of $\lambda_1 + \mu_1$, $\lambda_2 + \mu_2$, $\lambda_3 + \mu_3$, etc. since the real duration of each transition has an exponential distribution with a mean of $(\lambda_n + \mu_n)^{-1}$ (where n is the current state), and these are independent of each other.

Also note the stationary probabilities, say s_i, of the EMC are *not* the same as the real-time stationary probabilities p_i of the original Markov process. To make the corresponding conversion, we must take into consideration that the average time spent in State i is $\frac{1}{\mu_i + \lambda_i}$ (different, in general, for each state). The s_i probabilities thus need to be weighed by the average times as follows:

$$p_i = \left(\frac{s_i}{\mu_i + \lambda_i} \right) \div \left(\sum_{i=0}^{\infty} \frac{s_i}{\mu_i + \lambda_i} \right).$$

EMCs can help us with questions such as finding the expected number of visits to a state before absorption and with the following important issue.

8.4 PROBABILITY OF ULTIMATE ABSORPTION

Denoting a_n to be the probability of ultimate absorption if the current state is n, we have

$$a_n = \frac{\lambda_n}{\lambda_n + \mu_n} a_{n+1} + \frac{\mu_n}{\lambda_n + \mu_n} a_{n-1}$$

(depending on whether, in the next transition, we go up or down), where $n \geq 1$ and $a_0 = 1$. To solve, uniquely, this set of difference equations, we need yet another initial (or boundary) condition; we supply this in Proposition 8.2. Please note: even though these probabilities are being derived using the corresponding EMC, they apply to the original Markov process as well.

To solve for a_n, we first introduce $d_n = a_n - a_{n+1}$. The preceding equation can be rewritten as

$$(\lambda_n + \mu_n)a_n = \lambda_n a_{n+1} + \mu_n a_{n-1},$$

which further implies

$$a_n - a_{n+1} = \frac{\mu_n}{\lambda_n}(a_{n-1} - a_n)$$

or

$$d_n = \frac{\mu_n}{\lambda_n} d_{n-1}$$

for $n \geq 1$. The solution to these is easy to construct:

$$d_n = d_0 \prod_{i=1}^{n} \frac{\mu_i}{\lambda_i}.$$

One can show the sum of all d_n values must be equal to 1 (this is equivalent to $\lim_{n\to\infty} a_n = 0$; see subsequent proof), which means

$$d_n = \frac{\prod\limits_{i=1}^{n} \frac{\mu_i}{\lambda_i}}{1 + \sum\limits_{n=1}^{\infty} \left(\prod\limits_{i=1}^{n} \frac{\mu_i}{\lambda_i} \right)} \qquad (8.1)$$

and

$$a_m = \sum_{n=m}^{\infty} d_n = 1 - \sum_{n=0}^{m-1} d_n.$$

If the sum in the denominator of (8.1) *diverges*, the probability of ultimate extinction is 1, regardless of the initial state.

Proposition 8.2. $\sum_{i=0}^{\infty} d_n$ *can only have a value 0 or 1.*

Proof. Assume a_n is a nonincreasing sequence (i.e., $a_{n+1} \le a_n$) such that $\lim_{n\to\infty} a_n = a_\infty > 0$. For any n, $1 - a_n$ (probability of *escaping ultimate* extinction, starting from State n) cannot be greater than $1 - a_\infty$ for all n.

Let us denote by $b_{n,M}$ the probability of escaping extinction after exactly M transitions (starting in n). Since $b_{n,M}$ is also nonincreasing (i.e., $b_{n,M+1} < b_{n,M}$) $1 - a_n \le 1 - a_\infty$, there is m_1 such that $b_{n,m_1} \le 1 - \frac{a_\infty}{2}$. After these m_1 transitions we are in a state, say j, that must be in the range $[n, n + m_1]$. For any such $j \in [n,\ n + m_1]$

$$b_{j,M} \xrightarrow{M\to\infty} 1 - a_j \le 1 - a_\infty.$$

We can thus find m_2 such that $b_{j,m_2} \le 1 - \frac{a_\infty}{2}$ for each j.

Moreover,

$$b_{n,m_1+m_2} \le \left(1 - \frac{a_\infty}{2}\right) \cdot \sum_j p_j b_{j,m_2} \le \left(1 - \frac{a_\infty}{2}\right)^2,$$

where p_j is the probability of being in State j after m_1 transitions (starting in n). Repeating this argument indefinitely we get

$$b_{n,m_1+m_2+m_3+\cdots} \le \lim_{k\to\infty} \left(1 - \frac{a_\infty}{2}\right)^k = 0,$$

implying $\lim_{M\to\infty} b_{n,M} = 0$ for each n, and $a_\infty = 1$. Thus, a_∞ can have only two values: 0 or 1.   □

Example 8.3. Linear growth *without* immigration has $\lambda_n = n \cdot \lambda$ and $\mu_n = n \cdot \mu$. Thus, we get

$$d_n = \frac{\tau^n}{1 + \sum\limits_{k=1}^{\infty} \tau^k} = (1 - \tau)\tau^n,$$

where $\tau = \frac{\mu}{\lambda} < 1$, and

$$a_m = \sum_{n=m}^{\infty} d_n = \tau^m$$

(when $\tau \geq 1$, $a_m = 1$ and extinction becomes certain). This agrees with our old results. $\diamond$

Example 8.4. Consider a B&D process with $\lambda_n = \frac{6n^2}{1+0.3n^2}$ and $\mu_n = \frac{6n}{1+0.3n}$ (note State 0 is absorbing). Compute numerically the probability of ultimate absorption, given the process starts in State i (display these probabilities as a function of i).

Solution.

> $\lambda := n \rightarrow \dfrac{6 \cdot n^2}{1 + 0.3 \cdot n^2}$:

> $\mu := n \rightarrow \dfrac{6 \cdot n}{1 + 0.3 \cdot n}$:

> $S := \text{Re}\left(\displaystyle\sum_{m=0}^{\infty} \prod_{k=1}^{m} \dfrac{\mu(k)}{\lambda(k)}\right)$;

{This time, the infinite sum has an analytic solution.}

$$S := 4.2821$$

> *pointplot* $\left(\left[seq\left(\left[n, 1 - \text{Re}\left(\displaystyle\sum_{m=0}^{n-1} \dfrac{1}{S} \cdot \prod_{k=1}^{m} \dfrac{\mu(k)}{\lambda(k)}\right)\right], n = 0..15\right)\right]\right)$;

$\square$

8.5 MEAN TIME TILL ABSORPTION

When absorption is *certain*, it is interesting to investigate the expected time till absorption, say ω_n, given the process starts in State n. Considering what can happen in the next transition, we get

$$\omega_n = \frac{\lambda_n}{\lambda_n + \mu_n}\omega_{n+1} + \frac{\mu_n}{\lambda_n + \mu_n}\omega_{n-1} + \frac{1}{\lambda_n + \mu_n} \tag{8.2}$$

(the last term is the expected length of time till the next transition).

This implies

$$\omega_{n+1} = \left(1 + \frac{\mu_n}{\lambda_n}\right)\omega_n - \frac{\mu_n}{\lambda_n}\omega_{n-1} - \frac{1}{\lambda_n}.$$

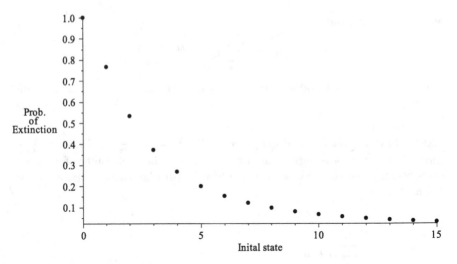

Similarly to the d_n of the previous section, we now introduce $\delta_n = \omega_n - \omega_{n+1}$ and rewrite the last equation as

$$\delta_n = \frac{\mu_n}{\lambda_n}\delta_{n-1} + \frac{1}{\lambda_n},$$

which yields

$$\delta_1 = \frac{\mu_1}{\lambda_1}\delta_0 + \frac{1}{\lambda_1}$$

$$\delta_2 = \frac{\mu_2}{\lambda_2}\delta_1 + \frac{1}{\lambda_2} = \frac{\mu_2\mu_1}{\lambda_2\lambda_1}\delta_0 + \frac{\mu_2}{\lambda_2\lambda_1} + \frac{1}{\lambda_2}$$

$$\delta_3 = \frac{\mu_3}{\lambda_3}\delta_2 + \frac{1}{\lambda_3} = \frac{\mu_3\mu_2\mu_1}{\lambda_3\lambda_2\lambda_1}\delta_0 + \frac{\mu_3\mu_2}{\lambda_3\lambda_2\lambda_1} + \frac{\mu_3}{\lambda_3\lambda_2} + \frac{1}{\lambda_3}$$

$$\vdots$$

One can show that, in general, $\delta_n \to 0$ as $n \to \infty$, implying

$$\omega_1 = \frac{1}{\mu_1} + \frac{\lambda_1}{\mu_1\mu_2} + \frac{\lambda_1\lambda_2}{\mu_1\mu_2\mu_3} + \frac{\lambda_1\lambda_2\lambda_3}{\mu_1\mu_2\mu_3\mu_4} + \cdots.$$

The remaining ω_n can be computed, recursively, based on (8.2) (note $\omega_0 = 0$).

Example 8.5. Linear growth without immigration.

$$\omega_1 = \frac{1}{\lambda}\sum_{i=1}^{\infty}\frac{\rho^i}{i} = -\frac{\ln(1-\rho)}{\lambda},$$

$$\omega_2 = (1 + \tau)\omega_1 - \frac{1}{\lambda},$$

$$\omega_3 = (1 + \tau)\omega_2 - \tau\omega_1 - \frac{1}{2\lambda}$$

$$= (1 + \tau + \tau^2)\omega_1 - \frac{1 + \tau}{\lambda} - \frac{1}{2\lambda},$$

$$\omega_4 = (1 + \tau)\omega_3 - \tau\omega_2 - \frac{1}{3\lambda}$$

$$= (1 + \tau + \tau^2 + \tau^3)\omega_1 - \frac{1 + \tau + \tau^2}{\lambda} - \frac{1 + \tau}{2\lambda} - \frac{1}{3\lambda}$$

$$\vdots$$

(note the pattern), where $\rho = \frac{\lambda}{\mu}$ and $\tau = \frac{\mu}{\tau}$. $\diamond$

Example 8.6. Consider a B&D process with

$$\lambda_n = \frac{6n}{1 + 0.3n},$$

$$\mu_n = \frac{6n^2}{1 + 0.3n^2}.$$

Verify absorption is certain (regardless of the initial state i), and find and display the mean time till absorption as a function of i.

Solution.

{Instruct MAPLE to do all calculations using 25-digit accuracy; this is necessary because our difference equations are numerically ill-conditioned.}

```
> Digits := 25 :
```

$$> \lambda := n \to \frac{6 \cdot n}{1 + 0.3 \cdot n} :$$

$$> \mu := n \to \frac{6 \cdot n^2}{1 + 0.3 \cdot n} :$$

```
> w_0 := 0 :
```

$$> w_1 := \mathrm{Re}\left(\frac{1}{\mu(1)} \cdot \sum_{m=0}^{\infty} \prod_{k=1}^{m} \frac{\lambda(k)}{\mu(k+1)}\right);$$

$$w_1 := 0.3056$$

```
> for n from 1 to 19 do
```

$$w_{n+1} := \frac{(\lambda(n) + \mu(n)) \cdot w_n - \mu(n) \cdot w_{n-1} - 1}{\lambda(n)};$$

end do:
> *pointplot* ([*seq* ([*n*, *w_n*], *n* = 0..20)]) ;

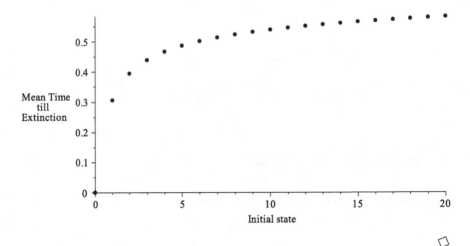

EXERCISES

Exercise 8.1. Consider an *M/M/*1queue with 17.3 arrivals per hour (on average), a mean service time of 4 min 26 s, and a probability that an arrival joins the system given by 0.62^k, where k is the number of customers present (counting the one being served). Find:

(a) The server utilization factor;
(b) The percentage of lost customers;
(c) The average size of the actual queue;
(d) The percentage of time with more than two customers waiting for service.

Exercise 8.2. Consider an *M/M/*1 queue with customers arriving at a rate of 12 per hour and an average service time of 10 min. Also, a customer who arrives and finds k people waiting for service walks away with a probability of $\frac{2k}{2k+1}$. Determine the stationary distribution of this process. What percentage of time will the server be idle in the long run?

Exercise 8.3. Consider another B&D process, with rates given by

$$\lambda_n = n\,\lambda \text{ for } n \geq 0,$$

$$\mu_n = \begin{cases} 0 & \text{when } n = 0, \\ \mu & \text{when } n \geq 1. \end{cases}$$

Find an expression for the probability of ultimate extinction, assuming that the process starts in State 3.

Exercise 8.4. Consider an $M/M/4$ queue with customers arriving at an average rate of 7.1 per hour and service time taking, on average, 25 min. Find the long-run

(a) Server utilization factor;
(b) Average number of customers waiting for service;
(c) Average waiting time;
(d) Percentage of time with no line.

Exercise 8.5. Consider an $M/M/1$ queue with 16.3 arrivals per hour (on average), a mean service time of 3 min 26 s, and a probability of an arrival joining the system of 0.65^k, where k is the number of customers *waiting*. Find the long-run:

(a) Server utilization factor;
(b) Percentage of lost customers (when a customer arrives, the probability that there will be n customers in the system is p_n – a nontrivial but true fact).
(c) The average number of customers waiting for service;
(d) The average waiting time.

Exercise 8.6. Consider a B&D process with the following (per-minute) rates:

$$\lambda_n = 0.69 \ln(1 + n),$$
$$\mu_n = \frac{3.2 \, n^{1.05}}{1 + n}.$$

Given the process is now in State 10, find the probability that it will become (sooner or later) trapped in State 0 (note State 0 is absorbing). If this probability is equal to 1, find the expected time till absorption (starting in State 10).

Exercise 8.7. Consider a B&D process with the following (per-minute) rates:

$$\lambda_n = 0.6\sqrt{n},$$
$$\mu_n = \frac{3n}{1 + n}.$$

Given that the process is now in State 30, find the probability that:

(a) It will become extinct (reaching State 0);
(b) The next three transitions will be all births;
(c) No transition will take place during the next 22 s.

CHAPTER 9
Continuous-Time Markov Chains

We generalize the processes discussed in the last three chapters even further by allowing an instantaneous transition from any state into any other (meaning *different*) state, with each such transition having its own specific rate. These new processes are so much more difficult to investigate that we are forced to abandon the infinite-state space and assume a finite (and usually quite small) number of states. Furthermore, the states no longer need to be integers (arbitrary labels will do since "bigger" and "smaller" no longer apply); integers, of course, may still be used as convenient labels.

9.1 BASICS

We will call these *general* Markov processes, that is, those that can jump from one state directly to any other state (not necessarily adjacent to it), continuous-time Markov chains (CTMCs).

Any such process is completely specified by the corresponding rates, say a_{ij}, of the process jumping directly from state i to state j. Remember this implies the probability that such a jump will occur during a brief time interval of length Δ is given by $a_{ij}\Delta + o(\Delta)$.

This time, we get the following set of difference-differential equations for $P_{i,n}(t)$:

$$\dot{P}_{i,n}(t) = \sum_{k \neq n} P_{i,k}(t) \cdot a_{k,n} - P_{i,n}(t) \cdot \sum_{j \neq n} a_{n,j},$$

which can be rewritten in the matrix form

$$\dot{\mathbb{P}}(t) = \mathbb{P}(t)\,\mathbb{A}$$

J. Vrbik and P. Vrbik, *Informal Introduction to Stochastic Processes with Maple*, Universitext, DOI 10.1007/978-1-4614-4057-4_9,
© Springer Science+Business Media, LLC 2013

assuming we define $a_{k,k} = -\sum_{j \neq k} a_{k,j}$ (the overall rate of *leaving* State k). These are called KOLMOGOROV'S FORWARD EQUATIONS, whereas $\mathbb{A}$ is the INFINITESIMAL GENERATOR of the process.

Similarly, one can derive Kolmogorov's *backward* equations:

$$\dot{\mathbb{P}}(t) = \mathbb{A}\,\mathbb{P}(t).$$

When the number of states is *finite*, the two sets of differential equations are, for practical purposes, equivalent.

Example 9.1. A machine can fail in two possible ways (e.g., mechanically, called state 1, or electronically, called state 2) at a rate of θ_1 and θ_2, respectively. Repairing a failure takes an exponential amount of time, with a mean value of μ_1 (mechanical failure) or μ_2 (electronic failure). If its fully operational state is labeled 0, then the infinitesimal generator of the process is

$$\mathbb{A} = \begin{array}{c} \\ 0 \\ 1 \\ 2 \end{array} \begin{array}{ccc} 0 & 1 & 2 \\ \hline -\theta_1 - \theta_2 & \theta_1 & \theta_2 \\ \frac{1}{\mu_1} & -\frac{1}{\mu_1} & 0 \\ \frac{1}{\mu_2} & 0 & -\frac{1}{\mu_2} \end{array}$$

$\diamondsuit$

We would like to learn how to solve Kolmogorov's equations using $\mathbb{P}(0) = \mathbb{I}$ as the initial condition (at time 0, $P_{i,n}$ is equal to $\delta_{i,n}$). Symbolically, the solution is given by

$$\mathbb{P}(t) = \exp(\mathbb{A}\,t),$$

in exact analogy with the familiar $y' = a \cdot y$ subject to $y(0) = 1$. Evaluating $\exp(\mathbb{A}\,t)$ *directly* (i.e., based on its Taylor expansion) would require us to find the limit of the following series:

$$\mathbb{I} + \mathbb{A}\,t + \mathbb{A}^2 \frac{t^2}{2!} + \mathbb{A}^3 \frac{t^3}{3!} + \mathbb{A}^4 \frac{t^4}{4!} + \cdots,$$

which (despite being an obvious solution to Kolmogorov's differential equation) is not very practical. Fortunately, there is an easier way of evaluating a function of a square matrix given by

$$f(\mathbb{A}) = \mathbb{C}_1 f(\omega_1) + \mathbb{C}_2 f(\omega_2) + \cdots + \mathbb{C}_N \, f(\omega_N),$$

where $\mathbb{C}_1, \mathbb{C}_2, \ldots, \mathbb{C}_N$ are the so-called CONSTITUENT MATRICES of $\mathbb{A}$ (N is the matrix size) and $\omega_1, \omega_2, \ldots, \omega_N$ are the corresponding EIGENVALUES of $\mathbb{A}$ (for the time being, they are assumed to be distinct). In our case, one of them must be equal to zero (why?). A method for finding constituent matrices of any $\mathbb{A}$ is discussed in Appendix 9.A. Fortunately, MAPLE has a built-in function for evaluating $\exp(\mathbb{A}t)$: `>MatrixExponential(A,t)`. Nevertheless (as with all other features of MAPLE used in this book), the reader should have a basic understanding of the underlying mathematics.

To generate a realization of any such process, one can show (in a manner similar to Sect. 7.1) that, while in State i, the time till the next transition is exponentially distributed with the mean of $\left(\sum_{j \neq i} a_{ij}\right)^{-1}$, while the conditional probability of entering State j is

$$\frac{a_{ij}}{\sum_{j \neq i} a_{ij}}$$

(algebraically independent of the transition time).

Example 9.2. Generate a random infinitesimal generator of a CTMC with seven states, labeled 1 to 7. Compute and plot $\Pr(X(t) = 5 \mid X(0) = 3)$.

Also, simulate one realization of this process (for 20 units of time) starting in State 3.

Solution.

```
> A := RandomMatrix (7, 7, generator = 0..0.43);
> for i from 1 to 7 do
      A_{i,i} := 0;
      A_{i,i} := -add (A_{i,j}, j = 1..7);
   end do:
> assume (t, real):
> Pr_{35} := Re (MatrixExponential(A, t)_{3,5}):
> plot (Pr_{35}, t = 0..4);
```

```
> (i, T) := (3, 20):
```
{specify initial state, and time T}
```
> (tt, n) := (0, i):
> for j from 1 while tt < T and A_{n,n} < 0 do
      tt := tt + Sample ( Exponential (-\frac{1}{A_{n,n}}), 1 )_1;
```

$$aux := \left[seq\left(\frac{\mathbb{A}_{n,k}}{-\mathbb{A}_{n,n}}, k = 1..7 \right) \right];$$

$aux_n := 0;$

$n := Sample\,(ProbabilityTable(aux), 1)_1 :$

$n := \mathrm{trunc}(n);$

$cond_j := t < tt, n;$

end do:

$> conditions := seq\,(cond_i, i = 1..j - 1) :$

$> f := piecewise(conditions):$

$> plot(f, t = 0..T, numpoints = 500);$

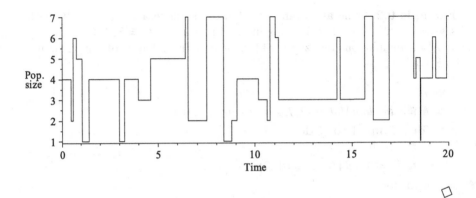

9.2 LONG-RUN PROPERTIES

We will now turn our attention to investigating properties of a CTMC when $t \to \infty$. To find these, we can bypass taking the limit of the complete $\mathbb{P}(t)$ solution but instead use a few shortcuts similar to those of the previous section.

STATIONARY PROBABILITIES

Let us first consider a case *without absorbing states* where the main issue is to find the *stationary probabilities*. This means that, in the $\dot{\mathbb{P}} = \mathbb{A}\,\mathbb{P}$ equation, $\dot{\mathbb{P}}$ becomes the zero matrix, and $\mathbb{P}$ itself will consist of *identical* rows (at a distant time, the solution no longer depends on the initial value), each equal to the vector of stationary probabilities, say s^T. We thus get

$$0 = A \begin{bmatrix} s^T \\ s^T \\ \vdots \\ s^T \end{bmatrix}$$

or, pulling out any single row (they are all identical),

$$0^T = As^T,$$

where 0 is the zero vector. Taking the transpose of the previous equation yields

$$A^Ts = 0.$$

Since $\det(A) = 0$, this system must have a *nonzero* solution that, after proper *normalization*, yields the stationary probabilities.

Example 9.3. Let

$$A = \begin{bmatrix} -8 & 3 & 5 \\ 2 & -6 & 4 \\ 1 & 6 & -7 \end{bmatrix}.$$

We thus must solve

$$\begin{bmatrix} -8 & 2 & 1 \\ 3 & -6 & 6 \\ 5 & 4 & -7 \end{bmatrix} \begin{bmatrix} s_1 \\ s_2 \\ s_3 \end{bmatrix} = \begin{bmatrix} 0 \\ 0 \\ 0 \end{bmatrix}.$$

Solution. We first set $s_3 = 1$ and delete the last equation (which must be a linear combination of the previous two), getting

$$\begin{bmatrix} -8 & 2 \\ 3 & -6 \end{bmatrix} \begin{bmatrix} s_1 \\ s_2 \end{bmatrix} = \begin{bmatrix} -1 \\ -6 \end{bmatrix}.$$

Solving these yields

$$\begin{bmatrix} s_1 \\ s_2 \end{bmatrix} = \frac{\begin{bmatrix} -6 & -2 \\ -3 & -8 \end{bmatrix} \begin{bmatrix} -1 \\ -6 \end{bmatrix}}{42} = \begin{bmatrix} \frac{18}{42} \\ \frac{51}{42} \end{bmatrix},$$

giving us $s^T = [\frac{18}{42} \; \frac{51}{42} \; 1]$. This is not a *distribution* yet. But, if we realize the solution of a homogeneous equation can be multiplied (or divided) by an arbitrary constant, then the correct answer must be $s^T = [\frac{18}{111} \; \frac{51}{111} \; \frac{42}{111}]$. $\square$

ABSORPTION ISSUES

Let us now assume the process has one or more ABSORBING STATES, implying absorption is *certain*. To get the *expected time* till absorption (into *any* one of the absorbing states), given that we start in State i (let us denote this expected value by ω_i), we realize that, due to the total-probability formula (taking the process to the next transition), we have

$$\omega_i = \frac{\sum_{j \neq i} a_{ij} \omega_j}{\sum_{j \neq i} a_{ij}} + \frac{1}{\sum_{j \neq i} a_{ij}}$$

whenever i is a nonabsorbing (transient) state. This is equivalent to

$$\omega_i \sum_{j \neq i} a_{ij} = \sum_{j \neq i} a_{ij} \omega_j + 1$$

or, in matrix form,

$$\widetilde{A}\omega = -1, \tag{9.1}$$

where $\widetilde{A}$ is the original infinitesimal generator *without* the absorbing-state *rows* (the equation does not hold for these) and without the absorbing-state *columns* (as the corresponding ω_k are all equal to zero). Solving this set of equations can be done easily with MAPLE.

Similarly, one can define τ_i as the second simple moment of TIME TILL ABSORPTION, say T, given the current state is i, and set up the following set of equations for these (based on $T^2 = T_0^2 + 2T_0T_1 + T_1^2$, where T_0 is the time till the next transition, and T_1 is the remaining time to get absorbed):

$$\tau_i = \frac{2}{\left(\sum_{j \neq i} a_{ij}\right)^2} + \frac{2 \sum_{j \neq i} a_{ij} \omega_j}{\left(\sum_{j \neq i} a_{ij}\right)^2} + \frac{\sum_{j \neq i} a_{ij} \tau_j}{\sum_{j \neq i} a_{ij}}$$

$$= \frac{2}{\left(\sum_{j \neq i} a_{ij}\right)^2} + \frac{2 \left(\omega_i \sum_{j \neq i} a_{ij} - 1\right)}{\left(\sum_{j \neq i} a_{ij}\right)^2} + \frac{\sum_{j \neq i} a_{ij} \tau_j}{\sum_{j \neq i} a_{ij}}$$

$$= \frac{2\omega_i}{\sum_{j \neq i} a_{ij}} + \frac{\sum_{j \neq i} a_{ij} \tau_j}{\sum_{j \neq i} a_{ij}},$$

implying

$$\widetilde{A}\tau = -2\omega. \tag{9.2}$$

Solving for τ_i and then subtracting ω_i^2 produces the corresponding variance. (One can obtain the same results more directly when the distribution of T is available.)

Example 9.4. Make State 7 of Example 9.2 absorbing. Find the distribution of the time till absorption, starting in State 3. Also, find the corresponding mean and standard deviation.

Solution.

```
> for i from  1 to  7 do
      A_{7,i} := 0 :
  end do:
> f := \frac{d}{dt} (Re (MatrixExponential(A, t))_{3,7}) :
> plot(f, t = 0..26);
```

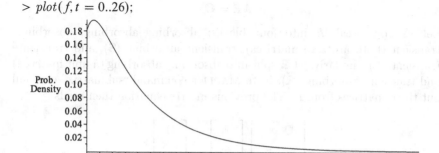

Time till absorption

```
> \mu := \int_0^\infty expand(t \cdot f) \, dt; {the mean}
```

$$\mu := 4.1074$$

```
> \omega := LinearSolve (SubMatrix(A, 1..6, 1..6), Vector(1..6, -1)) : \omega_3;
```
{verifying the corresponding formula}

$$4.1074$$

```
> \sqrt{\int_0^\infty expand ((t - \mu)^2 \cdot f) \, dt};
```

$$3.9415$$

```
> \sqrt{LinearSolve (SubMatrix(A, 1..6, 1..6), -2 \cdot \omega)_3 - \mu^2};
```

$$3.9415$$

{Verifying (9.2).}

$\square$

　　When there are at least two absorbing states, we would like to know the probabilities of ultimate absorption in each of these, given the process starts in State i. Denoting such probabilities r_{ik} (where i denotes the initial state and k the final, absorbing state), we can similarly argue

$$r_{ik} = \frac{\sum_{j \neq i} a_{ij} r_{jk}}{\sum_{j \neq i} a_{ij}}$$

or

$$\left(\sum_{j \neq i} a_{ij} \right) r_{ik} = \sum_{j \neq i} a_{ij} r_{jk}$$

whose matrix form is

$$\mathbb{A}\mathbb{R} = \mathbb{O}.$$

Let us now break $\mathbb{A}$ into four blocks: absorbing–absorbing, absorbing–transient (both are zero matrices), transient–absorbing ($\mathbb{S}$), and transient–transient ($\widetilde{\mathbb{A}}$); similarly, let $\mathbb{R}$ split into absorbing–absorbing (a unit matrix $\mathbb{I}$) and transient–absorbing ($\widetilde{\mathbb{R}}$). Note MAPLE's command "submatrix" can pull out these matrices from $\mathbb{A}$. The previous matrix equation then reads

$$\begin{bmatrix} \mathbb{O} & \mathbb{O} \\ \mathbb{S} & \widetilde{\mathbb{A}} \end{bmatrix} \begin{bmatrix} \mathbb{I} \\ \widetilde{\mathbb{R}} \end{bmatrix} = \begin{bmatrix} \mathbb{O} \\ \mathbb{O} \end{bmatrix},$$

which needs to be solved for

$$\widetilde{\mathbb{R}} = -\widetilde{\mathbb{A}}^{-1}\mathbb{S}. \tag{9.3}$$

This is consistent with what one would get using the embedded Markov chain approach of Sect. 8.3.

Example 9.5. Assuming we have a CTMC with five states (labeled 1 to 5 – the first and the last states are absorbing) and the following infinitesimal generator:

$$\mathbb{A} = \begin{bmatrix} 0 & 0 & 0 & 0 & 0 \\ 1 & -13 & 3 & 5 & 4 \\ 3 & 2 & -10 & 4 & 1 \\ 2 & 1 & 6 & -11 & 2 \\ 0 & 0 & 0 & 0 & 0 \end{bmatrix}$$

(all rates are per hour), we get

$$
\widetilde{\mathbb{R}} = -
\begin{bmatrix}
-13 & 3 & 5 \\
2 & -10 & 4 \\
1 & 6 & -11
\end{bmatrix}^{-1}
\begin{bmatrix}
1 & 4 \\
3 & 1 \\
2 & 2
\end{bmatrix}
=
\begin{bmatrix}
0.42903 & 0.57097 \\
0.60645 & 0.39355 \\
0.55161 & 0.44839
\end{bmatrix}.
$$

This means, for example, the probability of being absorbed by State 1, given that the process starts in State 4, is equal to 55.16%.

At the same time, the expected time till absorption (in either absorbing state) is

$$
-
\begin{bmatrix}
-13 & 3 & 5 \\
2 & -10 & 4 \\
1 & 6 & -11
\end{bmatrix}^{-1}
\begin{bmatrix}
1 \\
1 \\
1
\end{bmatrix}
=
\begin{bmatrix}
\frac{211}{930} \\
\frac{113}{465} \\
\frac{227}{930}
\end{bmatrix}.
$$

Starting in State 4, this yields $\frac{227}{930}$ h, or 14 min and 39 s. $\qquad\Diamond$

TABOO PROBABILITIES are those that add the condition of having to avoid a specific state (or a set of states). They are dealt with by making these "taboo" states absorbing.

Example 9.6. Returning to Example 9.2, find the probability of visiting State 1 *before* State 7 when starting in State 3.

Solution.

> $LinearSolve\,(SubMatrix(\mathbb{A}, 2..6, 2..6), -SubMatrix(\mathbb{A}, 2..6, [1, 7]))_{2,1}\,;$

$$0.3395$$

$\qquad\qquad\qquad\qquad\qquad\qquad\qquad\qquad\qquad\qquad\qquad\qquad\square$

Note a birth-and-death (B&D) process with finitely many states is just a special case of a CTMC process.

Example 9.7. Assume a B&D process has the following (per-hour) rates:

State	0	1	2	3	4	5	6
λ_n	3.2	4.3	4.0	3.7	3.4	2.8	0
μ_n	0	2.9	3.1	3.6	4.2	4.9	2.5

starting in State 5 at time 0.

1. Plot the probability of being in State 2 at time t.
2. Make State 0 absorbing, and plot the PDF of the time till absorption.

Solution.

$$
> \mathbb{A} := \begin{bmatrix}
0 & 3.2 & 0 & 0 & 0 & 0 & 0 \\
2.9 & 0 & 4.3 & 0 & 0 & 0 & 0 \\
0 & 3.1 & 0 & 4 & 0 & 0 & 0 \\
0 & 0 & 3.6 & 0 & 3.7 & 0 & 0 \\
0 & 0 & 0 & 4.2 & 0 & 3.4 & 0 \\
0 & 0 & 0 & 0 & 4.9 & 0 & 2.8 \\
0 & 0 & 0 & 0 & 0 & 2.5 & 0
\end{bmatrix} :
$$

> for i from 1 to 7 do
  $\mathbb{A}_{i,i} := -\sum_{j=1}^{7} \mathbb{A}_{i,j};$
  end do:

> $plot(MatrixExponential(\mathbb{A},t)_{6,3}, t = 0..4);$

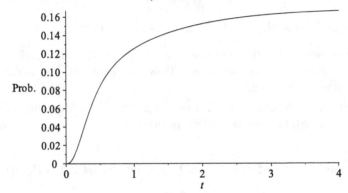

> $\mathbb{A}_{1,1} := 0 : \mathbb{A}_{1,2} := 0 :$
> $plot\left(\frac{d MatrixExponential(\mathbb{A},t)_{6,1}}{dt}, t = 0..30\right);$

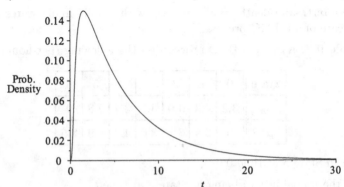

□

9.A FUNCTIONS OF SQUARE MATRICES

Functions of square matrices are defined by expanding a function in the usual, Taylor manner,

$$f(x) = c_0 + c_1 x + c_2 x^2 + c_3 x^3 + \cdots,$$

and replacing x with a square matrix $\mathbb{A}$. But we also know $\mathbb{A}$, when substituted for ω in the corresponding characteristic polynomial

$$\det(\omega \mathbb{I} - \mathbb{A}) = \omega^N - b_1 \omega^{N-1} + b_2 \omega^{N-2} - \cdots \pm b_N,$$

yields a zero matrix (this is the so-called Cayley–Hamilton theorem; $\mathbb{A}$ is assumed to be an $N \times N$ matrix, and b_j stands for the sum of its $j \times j$ *major* subdeterminants, that is, those that keep the same j rows as columns, for example, first, third, and sixth, deleting the rest – assuming $j = 3$).

This implies any power of $\mathbb{A}$ can be expressed as a linear combination of its first N powers (namely, $\mathbb{I}$, $\mathbb{A}$, $\mathbb{A}^2$, ..., $\mathbb{A}^{N-1}$), which further implies any power series in $\mathbb{A}$ (including the Taylor expansion above) can be reduced to a similar linear combination of these (finitely many) powers. To achieve that, we must first solve the recursive set of equations (see Appendix 3.A)

$$\mathbb{A}^n - b_1 \mathbb{A}^{n-1} + b_2 \mathbb{A}^{n-2} - \cdots \pm b_N \mathbb{A}^{n-N} = \mathbb{O}$$

for $\mathbb{A}^n$ by the usual technique of the *trial* solution

$$\mathbb{C} \cdot \omega^n$$

to discover $\mathbb{C}$ can be an arbitrary $N \times N$ matrix, as long as ω is a root of the original characteristic polynomial (i.e., it is an EIGENVALUE of $\mathbb{A}$). The fully general solution is then a linear combination of N such terms (assuming all N eigenvalues are distinct), namely,

$$\mathbb{A}^n = \sum_{j=1}^{N} \mathbb{C}_j \cdot \omega_j^n$$

(applying the SUPERPOSITION principle), where the $\mathbb{C}_j$ matrices are chosen to make this solution correct for the first N powers of $\mathbb{A}$ (thereby becoming the CONSTITUENT MATRICES of $\mathbb{A}$).

To evaluate $f(\mathbb{A})$, we now expand $f(x)$ into its Taylor series, say

$$f(x) = \sum_{n=0}^{\infty} f_n \cdot x^n,$$

replace x by $\mathbb{A}$, and then $\mathbb{A}^n$ by the preceding solution. This yields

$$f(\mathbb{A}) = \sum_{j=1}^{N} \mathbb{C}_j \sum_{n=0}^{\infty} f_n \cdot \omega_j^n = \sum_{j=1}^{N} f(\omega_j) \mathbb{C}_j.$$

One has to realize that both the eigenvalues and the corresponding constituent matrices may turn out to be *complex*, but since these must appear in *complex conjugate* pairs, the final result of $f(\mathbb{A})$ is always *real*.

Example 9.8. Consider the matrix

$$\mathbb{A} = \begin{bmatrix} 2 & 4 \\ 1 & -1 \end{bmatrix}.$$

The characteristic polynomial is $\omega^2 - \omega - 6$, and the eigenvalues are $\omega_1 = 3$ and $\omega_2 = -2$. Thus

$$\mathbb{C}_1 + \mathbb{C}_2 = \mathbb{I},$$
$$3\mathbb{C}_1 - 2\mathbb{C}_2 = \mathbb{A}.$$

Solution.

$$\mathbb{C}_1 = \frac{\mathbb{A} + 2\mathbb{I}}{5} = \begin{bmatrix} \frac{4}{5} & \frac{4}{5} \\ \frac{1}{5} & \frac{1}{5} \end{bmatrix},$$

$$\mathbb{C}_2 = \frac{3\mathbb{I} - \mathbb{A}}{5} = \begin{bmatrix} \frac{1}{5} & -\frac{4}{5} \\ -\frac{1}{5} & \frac{4}{5} \end{bmatrix}.$$

We can now evaluate any function of $\mathbb{A}$, for example,

$$e^{\mathbb{A}} = \begin{bmatrix} \frac{4}{5} & \frac{4}{5} \\ \frac{1}{5} & \frac{1}{5} \end{bmatrix} e^3 + \begin{bmatrix} \frac{1}{5} & -\frac{4}{5} \\ -\frac{1}{5} & \frac{4}{5} \end{bmatrix} e^{-2} = \begin{bmatrix} 16.095 & 15.96 \\ 3.99 & 4.125 \end{bmatrix}$$

and

$$\mathbb{A}^{-1} = \begin{bmatrix} \frac{4}{5} & \frac{4}{5} \\ \frac{1}{5} & \frac{1}{5} \end{bmatrix} /3 - \begin{bmatrix} \frac{1}{5} & -\frac{4}{5} \\ -\frac{1}{5} & \frac{4}{5} \end{bmatrix} /2 = \begin{bmatrix} \frac{1}{6} & \frac{2}{3} \\ \frac{1}{6} & -\frac{1}{3} \end{bmatrix}$$

(check).   □

Multiple Eigenvalues

We must now discuss what happens when we encounter double (triple, etc.) eigenvalues. We know such eigenvalues must satisfy not only the characteristic polynomial but also its first derivative (and its second derivative, in the case of triple eigenvalue, etc.). This indicates that there is yet another trial solution to the recursive set of equations for A^n, namely, $\mathbb{D} \cdot n \, \omega^{n-1}$ (and $\mathbb{E} \cdot n(n-1)\omega^{n-2}$ in the case of a triple roots...), where $\mathbb{D}$, $\mathbb{E}$, etc. are again arbitrary matrices. The consequence is the following modification of the previous formula: for multiple eigenvalues, we must also include terms of the $f'\left(\omega_j\right) \mathbb{D}_j$, $f''\left(\omega_j\right)\mathbb{E}_j$, etc. type (the total number of these is given by the multiplicity of ω_j).

To be a bit more explicit, suppose ω_1, ω_2, and ω_3 are identical. Then, instead of

$$f(\mathbb{A}) = f(\omega_1)\mathbb{C}_1 + f(\omega_2)\mathbb{C}_2 + f(\omega_3)\mathbb{C}_3 + \cdots,$$

we must use

$$f(\mathbb{A}) = f(\omega_1)\mathbb{C}_1 + f'(\omega_1)\mathbb{D}_1 + f''(\omega_1)\mathbb{E}_1 + \cdots.$$

To find the corresponding constituent matrices, we now use

$$\mathbb{C}_1 + \mathbb{C}_4 + \cdots + \mathbb{C}_n = \mathbb{I},$$
$$\omega_1\mathbb{C}_1 + \mathbb{D}_1 + \omega_4\mathbb{C}_4 + \cdots + \omega_n\mathbb{C}_n = \mathbb{A},$$
$$\omega_1^2\mathbb{C}_1 + 2\omega_2\mathbb{D}_1 + 2\mathbb{E}_1 + \omega_4^2\mathbb{C}_4 + \cdots + \omega_n^2\mathbb{C}_n = \mathbb{A}^2,$$

$$\vdots$$

$$\omega_1^{n-1}\mathbb{C}_1 + (n-1)\omega_2^{n-2}\mathbb{D}_1 + (n-1)(n-2)\omega_3^{n-3}\mathbb{E}_1$$
$$+\omega_4^{n-1}\mathbb{C}_4 + \cdots + \omega_n^{n-1}\mathbb{C}_n = \mathbb{A}^{n-1},$$

which are solved in exactly the same manner as before.

Example 9.9.

$$A = \begin{bmatrix} 4 & 7 & 2 \\ -2 & -2 & 0 \\ 1 & 5 & 4 \end{bmatrix}$$

and has a triple eigenvalue of 2. We get

$$\mathbb{C} = \mathbb{I},$$
$$2\mathbb{C} + \mathbb{D} = \mathbb{A},$$
$$4\mathbb{C} + 4\mathbb{D} + 2\mathbb{E} = \mathbb{A}^2,$$

which yields $\mathbb{C} = \mathbb{I}$,

$$\mathbb{D} = \mathbb{A} - 2\mathbb{I} = \begin{bmatrix} 2 & 7 & 2 \\ -2 & -4 & 0 \\ 1 & 5 & 2 \end{bmatrix},$$

and

$$\mathbb{E} = \frac{1}{2}\mathbb{A}^2 - 2\mathbb{D} - 2\mathbb{I}$$

$$= \frac{1}{2}\begin{bmatrix} 4 & 7 & 2 \\ -2 & -2 & 0 \\ 1 & 5 & 4 \end{bmatrix}^2 - 2\begin{bmatrix} 3 & 7 & 2 \\ -2 & -3 & 0 \\ 1 & 5 & 3 \end{bmatrix}$$

$$= \begin{bmatrix} -4 & -2 & 4 \\ 2 & 1 & -2 \\ -3 & -\frac{3}{2} & 3 \end{bmatrix}.$$

Thus,

$$\mathbb{A}^{-1} = \frac{1}{2}\begin{bmatrix} 1 & 0 & 0 \\ 0 & 1 & 0 \\ 0 & 0 & 1 \end{bmatrix} - \frac{1}{4}\begin{bmatrix} 2 & 7 & 2 \\ -2 & -4 & 0 \\ 1 & 5 & 2 \end{bmatrix} + \frac{1}{4}\begin{bmatrix} -4 & -2 & 4 \\ 2 & 1 & -2 \\ -3 & -\frac{3}{2} & 3 \end{bmatrix}$$

$$= \begin{bmatrix} -1 & -\frac{9}{4} & \frac{1}{2} \\ 1 & \frac{7}{4} & -\frac{1}{2} \\ -1 & -\frac{13}{8} & \frac{3}{4} \end{bmatrix}$$

(check).

APPLICATIONS

When dealing with a TCMC, the only function of $\mathbb{A}$ (which in this case must have at least one 0 eigenvalue) needed is

$$\exp(t\mathbb{A}),$$

where t is time. Note $t\mathbb{A}$ has the same constituent matrices as $\mathbb{A}$; similarly, the eigenvalues of $\mathbb{A}$ are simply multiplied by t to get the eigenvalues of $t\mathbb{A}$.

When a TCMC has one or more absorbing states, $\exp(t\mathbb{A})$ enables us to find the probabilities of absorption (before time t) and the distribution of the time till absorption (to do this, we must pool all absorbing states into one).

Example 9.10. Let

$$\mathbb{A} = \begin{bmatrix} 0 & 0 & 0 & 0 \\ 3 & -8 & 2 & 3 \\ 1 & 4 & -6 & 1 \\ 0 & 0 & 0 & 0 \end{bmatrix},$$

meaning the first and last states are absorbing. The eigenvalues are $0, 0, -4$, and -10. We thus get

$$f(\mathbb{A}) = f(0)\mathbb{C}_1 + f'(0)\mathbb{D}_1 + f(-4)\mathbb{C}_3 + f(-10)\mathbb{C}_4.$$

Using $f(x) = 1, x, x^2$, and x^3 yields

$$\mathbb{C}_1 + \mathbb{C}_3 + \mathbb{C}_4 = \mathbb{I},$$
$$\mathbb{D}_1 - 4\mathbb{C}_3 - 10\mathbb{C}_4 = \mathbb{A},$$
$$16\mathbb{C}_3 + 100\mathbb{C}_4 = \mathbb{A}^2,$$
$$-64\mathbb{C}_3 - 1000\mathbb{C}_4 = \mathbb{A}^3,$$

respectively. Solving this linear set of ordinary equations (ignoring the fact the unknowns are matrices) results in

$$\mathbb{C}_1 = \mathbb{I} - \tfrac{39}{400}\mathbb{A}^2 - \tfrac{7}{800}\mathbb{A}^3,$$
$$\mathbb{D}_1 = \mathbb{A} + \tfrac{7}{20}\mathbb{A}^2 + \tfrac{1}{40}\mathbb{A}^3,$$
$$\mathbb{C}_3 = \tfrac{5}{48}\mathbb{A}^2 + \tfrac{1}{96}\mathbb{A}^3,$$
$$\mathbb{C}_4 = -\tfrac{1}{150}\mathbb{A}^2 - \tfrac{2}{600}\mathbb{A}^3.$$

A routine evaluation then yields

$$\mathbb{C}_1 = \begin{bmatrix} 1 & 0 & 0 & 0 \\ \tfrac{1}{2} & 0 & 0 & \tfrac{1}{2} \\ \tfrac{1}{2} & 0 & 0 & \tfrac{1}{2} \\ 0 & 0 & 0 & 1 \end{bmatrix},$$

$$\mathbb{D}_1 = \mathbb{O},$$

$$\mathbb{C}_3 = \begin{bmatrix} 0 & 0 & 0 & 0 \\ -\frac{1}{3} & \frac{1}{3} & \frac{1}{3} & -\frac{1}{3} \\ -\frac{2}{3} & \frac{2}{3} & \frac{2}{3} & -\frac{2}{3} \\ 0 & 0 & 0 & 0 \end{bmatrix},$$

$$\mathbb{C}_4 = \begin{bmatrix} 0 & 0 & 0 & 0 \\ -\frac{1}{6} & \frac{2}{3} & -\frac{1}{3} & -\frac{1}{6} \\ \frac{1}{6} & -\frac{2}{3} & \frac{1}{3} & \frac{1}{6} \\ 0 & 0 & 0 & 0 \end{bmatrix}.$$

Thus, $\exp(\mathbb{A}t)$ is computed from

$$\mathbb{C}_1 + \mathbb{C}_3 e^{-4t} + \mathbb{C}_4 e^{-10t}.$$

The elements of the first and last columns of $\mathbb{C}_1$ (the limit of the previous expression when $t \to \infty$) yield the probabilities of ultimate absorption in the first (last) state, given the initial state.

The sum of the first and the last columns of $\exp(\mathbb{A}t)$, namely,

$$\begin{bmatrix} 1 \\ 1 - \frac{2}{3}e^{-4t} - \frac{1}{3}e^{-10t} \\ 1 - \frac{4}{3}e^{-4t} + \frac{1}{3}e^{-10t} \\ 1 \end{bmatrix},$$

provides the distribution function of the time till absorption (whether into the first or last state), given the initial state. Based on that, we can answer any probability question, find the corresponding mean and standard deviation, etc. $\diamondsuit$

Example 9.11. (Continuation of Example 9.10). Given we start in the second state, what is the probability that absorption will take more than 0.13 units of time?

Solution.

$$\tfrac{2}{3}e^{-0.52} + \tfrac{1}{3}e^{-1.3} = 48.72\%.$$

$\square$

Example 9.12. (Continuation of Example 9.10). Given we start in the third state, what is the expected time till absorption (regardless of where) and the corresponding standard deviation?

Solution.

$$\int_0^\infty t \left(\tfrac{16}{3}e^{-4t} - \tfrac{10}{3}e^{-10t}\right) dt = 0.3$$

$$\sqrt{\int_0^\infty (t - 0.3)^2 \left(\tfrac{16}{3}e^{-4t} - \tfrac{10}{3}e^{-10t}\right) dt} = 0.2646$$

Again, the reader should verify that these agree with the algebraic solution given by (9.1) and (9.2).   ◻

EXERCISES

For the following questions, a • denotes rates that are meaningless and thus unnecessary to define.

Exercise 9.1. Assume a B&D process has the following (per-hour) rates

State	0	1	2	3	4	5	6
λ_n	0	4.3	4.0	3.7	3.1	2.8	0
μ_n	0	2.9	3.1	3.6	4.2	4.9	0

starting in State 2 at time 0.

(a) Find the expected time till absorption (to either absorbing state).
(b) What is the probability that the process will end up in State 6?
(c) What is the probability that the process will end up in State 6, never having visited State 1?

Exercise 9.2. Consider a time-continuous Markov process with four states (called first, second, third, and last) and the following (per-hour) rates of individual transitions:

$$\begin{bmatrix} \bullet & 3.2 & 1.3 & 0.4 \\ 2.0 & \bullet & 1.8 & 0.7 \\ 2.3 & 1.5 & \bullet & 3.0 \\ 1.9 & 2.1 & 0.9 & \bullet \end{bmatrix}.$$

(a) Find the (exact) corresponding stationary probabilities. How often, on average, is the first state visited (in the long run)?
(b) Now the process is in the second state. What is the probability that it will be in the last state 7 min later?
(c) Now the process is in the second state. What is the probability that it will be in the last state 7 min later, without ever (during those 7 min) having visited the third state?
(d) Now the process is in the second state. Find the expected time till the process enters (for the first time from now) the third state and the corresponding standard deviation.

Exercise 9.3. Evaluate $\ln(3\mathbb{I} - \mathbb{A})$, where

(a) $\mathbb{A} = \begin{bmatrix} 2 & -3 & 0 & 4 \\ 3 & 1 & -4 & 3 \\ -2 & 5 & 3 & 0 \\ 2 & 2 & -1 & -5 \end{bmatrix}$,

(b) $\mathbb{A} = \begin{bmatrix} 7 & -15 & 10 \\ 0 & 2 & 0 \\ -5 & 15 & -8 \end{bmatrix}$.

Exercise 9.4. Consider a time-continuous Markov process with five states and the following (per-hour) rates of individual transitions:

$$\begin{bmatrix} \bullet & 0 & 0 & 0 & 0 \\ 2.0 & \bullet & 1.8 & 0.7 & 1.3 \\ 2.3 & 1.5 & \bullet & 3.0 & 0.8 \\ 1.9 & 2.1 & 0.9 & \bullet & 1.7 \\ 0 & 0 & 0 & 0 & \bullet \end{bmatrix}$$

(indicating that the first and last states are absorbing). Given that now the process is in the third (middle) state, find:

(a) The expected time till absorption (in either absorbing state) and the corresponding standard deviation;
(b) The *exact* probability of getting absorbed in the last state.

Exercise 9.5. Consider a CTMC with the following (per-hour) rates:

$$
\begin{bmatrix}
\bullet & 0 & 0 & 0 & 0 & 0 \\
0.9 & \bullet & 2.3 & 1.8 & 2.0 & 0 \\
0 & 1.1 & \bullet & 2.0 & 1.7 & 1.4 \\
4.1 & 0 & 2.7 & \bullet & 0.4 & 2.8 \\
2.1 & 3.2 & 0 & 0.9 & \bullet & 3.1 \\
0 & 0 & 0 & 0 & 0 & \bullet
\end{bmatrix}
$$

(States 1 and 6 are clearly absorbing). Given the process starts in State 3, compute:

(a) The probability that it will end up in one of the absorbing states within the next 23 min;

(b) The expected time till absorption and the corresponding standard sdeviation;

(c) The exact (i.e., fractional) probability of being absorbed, sooner or later, by State 6.

CHAPTER 10
Brownian Motion

Empirically, BROWNIAN MOTION was discovered by the biologist Robert Brown, who observed, and was puzzled by, microscopic movement of tiny particles suspended in water (at that time, the reason for this motion was not yet understood). Due to the irregular thermal motion of individual water molecules, each such particle will be pushed around, following an irregular path in all three dimensions. We study only the single-particle, one-dimensional version of this phenomenon. To make the issue more interesting, we often assume the existence of an absorbing state (or a BARRIER) that, when reached, terminates the particle's motion.

10.1 BASICS

Brownian motion is our only example of a process with both time and state space being *continuous*. There are two alternate names under which the process is known: statisticians know it as a WIENER PROCESS, whereas physicists call it DIFFUSION.

If we restrict ourselves to a pointlike particle that moves only in *one* spatial dimension, we can visualize its movement as a limit of the process in which one bets \$1 on a flip of a coin, $Y(n)$ being the total net win (loss) after n rounds of the game. Defining $X(t) = \sqrt{\Delta} \cdot Y(n)$, where $t = \Delta \cdot n$, and letting Δ approach zero while correspondingly increasing the value of n, yields a one-dimensional Brownian motion.

Mathematically, the process can be introduced via the following postulates.

1. For each δ

$$\lim_{s \to 0} \frac{\Pr\left(|X(t+s) - X(t)| \geq \delta\right)}{s} = 0,$$

J. Vrbik and P. Vrbik, *Informal Introduction to Stochastic Processes with Maple*, Universitext, DOI 10.1007/978-1-4614-4057-4_10, © Springer Science+Business Media, LLC 2013

which means there are no instantaneous jumps – the process is continuous.

2. Each $X(t + s) - X(t)$ increment is (statistically) independent not only of the previous (PAST) values of X (Markovian property) but also of the current (PRESENT) value $X(t)$; furthermore, the distribution of any such increment will depend on s but must be (algebraically) independent of t – the distribution of these increments is *homogeneous* in time. This implies (based on central limit theorem) that $X(t + s) - X(t)$ is *normal*, with a mean of $d \cdot s$ and variance of $c \cdot s$ (both must be proportional to s), where d and c are the basic parameters of Brownian motion called DRIFT and DIFFUSION coefficients, respectively.

What follows is an example of what a realization of such a process (with $d = 0.1$ and $c = 5.3$) might look like.

```
> (aux, step) := (0, 0.1) :
> r := [0, aux] :
> for t from  0 to  100 by  step do
      aux := aux + Sample(RandomVariable(Normal(0.1 · step,
            √5.3 · step)), 1)₁ :
      r := r, [t, aux];
  end do:
> listplot([r]);
```

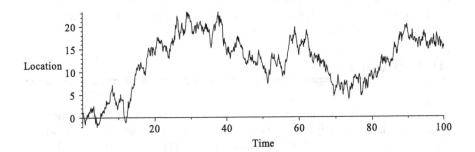

10.2 CASE OF $d = 0$

In this section we assume there is no drift, implying the process is equally likely to go up as it is to go down.

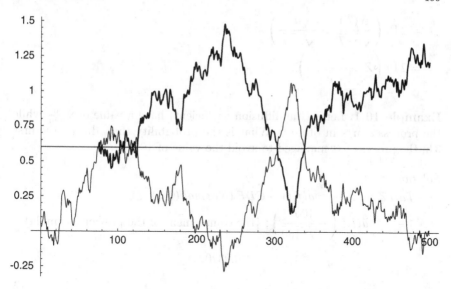

Fig. 10.1: Flip-over property

Reaching a Before Time T

The first question we will try to answer is this: Assuming the process starts in state 0 [i.e., $X(0) = 0$], what is the probability that it will attain a value of a, at least once, before time T?

Let us first visualize a possible realization of the process (a continuous path starting at the origin) that visits State a before time T. The path then continues until it terminates at $X(T)$. For every path that does this and ends in a state greater than a, there is an equally likely path (its "flip over" starting from the *first* visit to State a), which ends up in a state smaller than a (Fig. 10.1).

This argument can be reversed: for each path that has $X(T) > a$ there is an *equally likely path* that at one point must have also reached a but ended up in $X(T) < a$. The answer to the original question must therefore be equal to *double* the probability of $X(T) > a$, namely,

$$\Pr\left(\max_{0 \leq t \leq T} X(t) > a\right)$$

$$= \Pr\left(\max_{a \leq t \leq T} X(t) > a \cap X(t) > a\right) + \Pr\left(\max_{0 \leq t \leq T} X(t) > a \cap X(t) < a\right)$$

$$= 2\Pr\left(\max_{0 \leq t \leq T} X(t) > a \cap X(T) > a\right)$$

$$= 2\Pr\left(X(T) > a\right)$$

$$= 2\Pr\left(\frac{X(T)}{\sqrt{c \cdot T}} > \frac{a}{\sqrt{c \cdot T}}\right)$$

$$= 2\Pr\left(Z > \frac{a}{\sqrt{c \cdot T}}\right).$$

Example 10.1. Let c (the diffusion coefficient) have a value of $9\frac{cm^2}{min}$ while the process starts in State 0. What is the probability that, during the first 3 h, the process has managed to avoid the value of 100 cm?

Solution.

> $TailZ := z \to evalf\left(1 - CDF\left(Normal\left(0, 1\right), z\right)\right):$

> $1 - 2 \cdot TailZ\left(\dfrac{100}{\sqrt{9 \cdot 180}}\right)$; {the complement of the previous formula}

$$0.98702$$

$\square$

Two Extensions

1. When $a < 0$, consider the flip-over argument. The probability of reaching State a (at least once) before time T is thus

$$2\Pr\left(Z > \frac{|a|}{\sqrt{c \cdot T}}\right).$$

2. When $X(0) = x$, the probability of reaching State a before $t = T$ equals

$$2\Pr\left(Z > \frac{|a - x|}{\sqrt{c \cdot T}}\right),$$

based on *state-space* homogeneity.

Reaching y While Avoiding 0

A similar approach will help us answer yet another question: What is the probability that a Wiener process that starts in State $x > 0$ will be, at time T, in a state greater than y without ever dipping below 0 (we can visualize 0 as an absorbing state) – let us denote this probability $A(x, y, T)$.

We first consider all paths that start in State x and end up (at $t = T$) in a state $> y$. This set can be divided into two parts – those paths that managed to avoid State 0, and those that did not. Each of those paths that visited State 0 at least once has an equally likely counterpart (its 0 flip over, starting from the first visit to State 0) that ends up (at T) in a state lower than $-y$ (the reverse is also true, resulting in a one-to-one correspondence). The probability of $X(T) > y$ *and* of visiting 0 prior to $t = T$ is thus $\Pr(X(T) < -y \mid X(0) = x)$. We can put this more precisely as

$$\Pr\left(X(T) > y \mid X(0) = x\right)$$

$$= \Pr\left(X(T) > y \cap \min_{0 \le t \le T} X(t) > 0 \mid X(0) = x\right)$$

$$+ \Pr\left(X(T) > y \cap \min_{0 \le t \le T} X(t) < 0 \mid X(0) = x\right)$$

$$= A(x, y, T) + \Pr\left(X(T) < -y \cap \min_{0 \le t \le T} X(t) < 0 \mid X(0) = x\right)$$

$$= A(x, y, T) + \Pr\left(X(T) < -y \mid X(0) = x\right).$$

The final answer is therefore

$$A(x, y, T) = \Pr(X(T) > y \mid X(0) = x) - \Pr(X(T) < -y \mid X(0) = x)$$

$$= \Pr\left(Z > \frac{y - x}{\sqrt{c \cdot T}}\right) - \Pr\left(Z < \frac{-y - x}{\sqrt{c \cdot T}}\right)$$

$$= \Pr\left(Z > \frac{y - x}{\sqrt{c \cdot T}}\right) - \Pr\left(Z > \frac{y + x}{\sqrt{c \cdot T}}\right). \tag{10.1}$$

Example 10.2. Using the same $c = 9\frac{\text{cm}^2}{\text{min}}$ but a new initial value of $X(0) = 30$ cm, find the probability that the process will be, 3 h later, in a state higher than 70 cm, without ever visiting zero. Note if we make State 0 absorbing, then the same question is (more simply): $\Pr(X(3\,\text{h}) > 70\,\text{cm} \mid X(0) = 30\,\text{cm})$.

Solution.

$$> TailZ\left(\frac{40}{\sqrt{9 \cdot 180}}\right) - TailZ\left(\frac{100}{\sqrt{9 \cdot 180}}\right);$$

$$0.15367$$

$\Box$

We can generalize (10.1) as follows. If the absorbing state is a (instead of 0) and both $x > a$ and $y > a$, then

$$\Pr\left(X(T) > y \cap \min_{0 \le t \le T} X(t) > a \mid X(t) = x\right)$$

$$= \Pr\left(Z > \frac{y - x}{\sqrt{c \cdot T}}\right) - \Pr\left(Z > \frac{y + x - 2a}{\sqrt{c \cdot T}}\right). \tag{10.2}$$

Similarly (when $x < a$ and $y < a$),

$$\Pr\left(X(t) < y \cap \max_{0 \le t \le T} X(t) < a \mid X(t) = x\right)$$

$$= \Pr\left(Z > \frac{x-y}{\sqrt{c \cdot T}}\right) - \Pr\left(Z > \frac{2a-y-x}{\sqrt{c \cdot T}}\right).$$

Furthermore, (10.2) implies (when both x and y are higher than a) that $X(T) = a$ (due to having been absorbed therein) with a (discrete) probability of $2\Pr\left(Z > \frac{x-a}{\sqrt{cT}}\right)$. The rest of the $X(t)$ distribution is provided by the PDF

$$f_{X(T)}(y) = \frac{\exp\left(-\frac{(y-x)^2}{2cT}\right)}{\sqrt{2\pi cT}} - \frac{\exp\left(-\frac{(y+x-2a)^2}{2cT}\right)}{\sqrt{2\pi cT}}.$$

Proposition 10.1. *Based on the preceding distribution, the expected value of $X(T)$ is x, for any T. (A rather surprising result: the expected value is the same with or without the absorbing barrier!)*

Proof.

$$\mathbb{E}\left(X(T)\right)$$

$$= 2\Pr\left(Z > \frac{x-a}{\sqrt{cT}}\right) \cdot a + \frac{\int_a^\infty y \exp\left(-\frac{(y-x)^2}{2cT}\right) dy}{\sqrt{2\pi cT}}$$

$$- \frac{\int_a^\infty y \exp\left(-\frac{(y+x-2a)^2}{2cT}\right) dy}{\sqrt{2\pi cT}}$$

$$= 2\Pr\left(Z > \frac{x-a}{\sqrt{cT}}\right) \cdot a + \frac{\int_a^\infty (y-x) \exp\left(-\frac{(y-x)^2}{2cT}\right) dy}{\sqrt{2\pi cT}}$$

$$- \frac{\int_a^\infty (y+x-2a) \exp\left(-\frac{(y+x-2a)^2}{2cT}\right) dy}{\sqrt{2\pi cT}} + x\frac{\int_a^\infty \exp\left(-\frac{(y-x)^2}{2cT}\right) dy}{\sqrt{2\pi cT}}$$

$$- (2a-x)\frac{\int_a^\infty \exp\left(-\frac{(y+x-2a)^2}{2cT}\right) dy}{\sqrt{2\pi cT}}$$

$$= 2\Pr\left(Z > \frac{x-a}{\sqrt{cT}}\right) \cdot a + \frac{\int_{a-x}^\infty u \exp\left(-\frac{u^2}{2cT}\right) du}{\sqrt{2\pi cT}} - \frac{\int_{x-a}^\infty u \exp\left(-\frac{u^2}{2cT}\right) du}{\sqrt{2\pi cT}}$$

$$+ x\Pr\left(Z > \frac{a-x}{\sqrt{cT}}\right) - (2a-x)\Pr\left(Z > \frac{x-a}{\sqrt{cT}}\right)$$

$$= \frac{\int_{a-x}^{x-a} u \exp\left(-\frac{u^2}{2cT}\right) du + \int_{x-a}^\infty u \exp\left(-\frac{u^2}{2cT}\right) du}{\sqrt{2\pi cT}}$$

$$-\frac{\int_{x-a}^{\infty} u \exp\left(-\frac{u^2}{2cT}\right) du}{\sqrt{2\pi cT}} + x \Pr\left(Z > \frac{a-x}{\sqrt{cT}}\right) + x \Pr\left(Z > \frac{x-a}{\sqrt{cT}}\right)$$

$$= x.$$

□

RETURNING TO 0

Finally, we tackle the following question: Given a process starts in State 0, what is the probability it will return to 0 (at least once) during a time interval $[t_0, t_1]$? The resulting formula will prove to be relatively simple, but deriving it is rather tricky and will be done in the following two parts.

DISTRIBUTION OF T_a

Let T_a be the time at which the process visits State a for the first time. We already know (that was our first issue)

$$\Pr(T_a < t) = 2\Pr(X(t) > |a|)$$

$$= 2\Pr\left(Z > \frac{|a|}{\sqrt{c \cdot t}}\right)$$

$$= \frac{2}{\sqrt{2\pi}} \int_{\frac{|a|}{\sqrt{c \cdot t}}}^{\infty} \exp\left(-\frac{z^2}{2}\right) dz.$$

This is, of course, the distribution function of T_a. Differentiating with respect to t yields the corresponding PDF:

$$f(t) = \frac{|a|}{\sqrt{2\pi c} \cdot t^{3/2}} \cdot \exp\left(-\frac{a^2}{2ct}\right), \tag{10.3}$$

where $t > 0$.

FINAL FORMULA

Let A be the event of visiting State 0 between time t_0 and t_1, and let x be the value of the process at time t_0, that is, $x = X(t_0)$. Using the (extended) formula of total probability, we get

$$\Pr(A \mid X(0) = 0) = \int_{-\infty}^{\infty} \Pr(A \mid X(t_0) = x) \cdot f_{X(t_0)}(x \mid X(0) = 0) \, dx.$$

Given the process is in State x at time t_0, we know that the PDF of the *remaining* time to return to State 0 is given by (10.1), with $|a| = |x|$. We can thus compute

$$\Pr(A \mid X(t_0) = x) = \int_0^{t_1-t_0} \frac{|x|}{\sqrt{2\pi c} \cdot t^{3/2}} \cdot \exp(-\tfrac{x^2}{2ct}) \, dt.$$

Since $X(t_0) \in \mathcal{N}(0, \sqrt{c \cdot t_0})$, we also know

$$f_X(x \mid X(0) = 0) = \frac{\exp\left(-\dfrac{x^2}{2ct_0}\right)}{\sqrt{2\pi c t_0}}.$$

The answer is provided by the following double integral:

$$\Pr(A \mid X(0) = 0) = \int_{-\infty}^{\infty} \frac{\exp(-\frac{x^2}{2ct_0})}{\sqrt{2\pi c t_0}} \int_0^{t_1-t_0} \frac{|x|}{\sqrt{2\pi c} \cdot t^{3/2}} \cdot \exp(-\tfrac{x^2}{2ct}) \, dt \, dx.$$

By performing the dx integration first,

$$\int_{-\infty}^{\infty} |x| \exp\left(-\frac{x^2}{2\sigma^2}\right) dx = 2 \int_0^{\infty} x \exp\left(-\frac{x^2}{2\sigma^2}\right) dx$$

$$= 2\sigma^2 \int_0^{\infty} \exp(-u) \, du$$

$$= 2\sigma^2,$$

we obtain

$$\Pr(A \mid X(0) = 0) = \frac{1}{\sqrt{2\pi c t_0}} \int_0^{t_1-t_0} \frac{2c}{\sqrt{2\pi c} \cdot t^{3/2}} \cdot \frac{t_0 t}{t_0 + t} \, dt$$

$$= \frac{1}{\pi} \int_0^{t_1-t_0} \frac{t_0}{t_0 + t} \sqrt{\frac{1}{t_0 t}} \, dt$$

$$= \frac{2}{\pi} \int_0^{\sqrt{\frac{t_1-t_0}{t_0}}} \frac{du}{1 + u^2}$$

$$= \frac{2}{\pi} \arctan \sqrt{\frac{t_1}{t_0} - 1},$$

where $u = \sqrt{\dfrac{t}{t_0}}$. Since $\arctan q \equiv \arccos \dfrac{1}{\sqrt{1+q^2}}$, the final formula can be expressed as

$$\frac{2}{\pi} \arccos \sqrt{\frac{t_0}{t_1}}.$$

Example 10.1. When $c = 3\frac{cm^2}{min}$, the probability of crossing 0 (at least once) between 1 and 2 h after we start the process (in State 0) is $\dfrac{2}{\pi} \arccos \sqrt{\frac{1}{2}} = 50\%$ (independent of the value of c).

10.3 DIFFUSION WITH DRIFT

When $d \neq 0$, the process will have a steady (linear) increasing or decreasing (depending on the sign of d) component called DRIFT (this is like betting on roulette rather than using a fair coin where our chances of winning a single round are slightly less than 50%). The main result now is that the distribution of $X(t)$, given $X(t_0) = x_0$ (the last previously observed value), is normal with a mean of $x_0 + d(t - t_0)$ and a variance of $c(t - t_0)$.

Example 10.2. Suppose $d = -4\frac{cm}{h}$ and $c = 3\frac{cm^2}{h}$. Find $\Pr(X(12:00) < 0)$, given that at 9:30 the process was 5 cm from the origin.

Solution.

$> 1 - TailZ\left(\dfrac{0-5-(-4)\cdot 2.5}{\sqrt{3\cdot 2.5}}\right);$

$$0.96605$$

$\Box$

In this context, we investigate one more issue: Assuming we know the values of both c and d, and given the process has been observed at only a few isolated points in time [say $X(t_1) = x_1,\ X(t_2) = x_2,\ X(t_3) = x_3,\ \ldots$], what is the conditional distribution of $X(t)$ somewhere in between?

The solution is simple in principle but messy in terms of the ensuing algebra. The main thing to realize is that, due to the Markovian property, only the most recent piece of information from the past matters. Similarly, we can ignore all information from the future (i.e., $t_i > t$), except the one closest to t. We can thus reduce the question to: find the conditional PDF of $X(t)$, given $X(t_1) = x_1$ and $X(t_2) = x_2$, assuming $t_1 < t < t_2$.

To find the conditional PDF of $X(t)$, given the values of the process at t_1 and t_2, we start with

$$\Pr(x \leq X(t) < x + \Delta \mid X(t_1) = x_1 \cap X(t_2) = x_2)$$
$$= \Pr(B \mid A \cap C)$$

$$= \frac{\Pr(B \cap A \cap C)}{\Pr(A \cap C)}$$

$$= \frac{\Pr(B \cap A \cap C)}{\Pr(A \cap B)} \cdot \frac{\Pr(A \cap B)}{\Pr(A)} \cdot \frac{\Pr(A)}{\Pr(A \cap C)}$$

$$= \frac{\Pr(C \mid A \cap B) \cdot \Pr(B \mid A)}{\Pr(C \mid A)}$$

(it was necessary to rearrange the conditional probabilities chronologically since so far we know only how to forecast forward in time).

When we divide by Δ and take the $\Delta \to 0$ limit (and utilizing the Markovian property), the last expression equals

$$\lim_{\Delta \to 0} \frac{\frac{\Pr(C \mid B)}{\Delta}}{\frac{\Pr(C \mid A)}{\Delta}} \cdot \frac{\Pr(B \mid A)}{\Delta} = \frac{f_{X(t_2)}(x_2 \mid X(t) = x)}{f_{X(t_2)}(x_2 \mid X(t_1) = x_1)} \cdot f_{X(t)}(x \mid X(t_1) = x_1).$$

We know each of the conditional PDFs is normal, with the mean and variance computed by multiplying each c and d (respectively) by the corresponding time increment. Thus we get, for the conditional PDF of $X(t)$,

$$\frac{\frac{1}{\sqrt{2\pi c(t_2-t)}} \exp\left(-\frac{(x_2-x-d(t_2-t))^2}{2c(t_2-t)}\right)}{\frac{1}{\sqrt{2\pi c(t_2-t_1)}} \exp\left(-\frac{(x_2-x_1-d(t_2-t_1))^2}{2c(t_2-t_1)}\right)}$$

$$\times \frac{1}{\sqrt{2\pi c(t-t_1)}} \exp\left(-\frac{(x-x_1-d(t-t_1))^2}{2c(t-t_1)}\right).$$

The rest is an issue of the following algebraic simplification:

For a *constant* multiplying the final $\exp(\cdots)$ we get, almost immediately, a value of

$$\frac{1}{\sqrt{2\pi \cdot \frac{(t-t_1)(t_2-t)}{t_2-t_1}}}.$$

For the denominator in $\exp\left(-\frac{\cdots}{2c}\right)$ we obtain

$$\frac{(x_2-x-d(t_2-t))^2}{t_2-t} + \frac{(x-x_1-d(t-t_1))^2}{t-t_1} - \frac{(x_2-x_1-d(t_2-t_1))^2}{t_2-t_1}.$$

To simplify this expression, we first collect terms proportional to d^2:

$$\frac{(t_2-t)^2}{t_2-t} + \frac{(t-t_1)^2}{t-t_1} - \frac{(t_2-t_1)^2}{t_2-t_1} = (t_2-t) + (t-t_1) - (t_2-t_1) = 0,$$

then those proportional to d:

$$\frac{2(x_2-x)(t_2-t)}{t_2-t} + \frac{2(x-x_1)(t-t_1)}{t-t_1} - \frac{2(x_2-x_1)(t_2-t_1)}{t_2-t_1} = 0.$$

We see the answer will be free of the drift parameter d.

Finally, collecting the remaining terms yields

$$\frac{(x_2 - x)^2}{t_2 - t} + \frac{(x - x_1)^2}{t - t_1} - \frac{(x_2 - x_1)^2}{t_2 - t_1} = \frac{(x(t_2 - t_1) - x_1(t_2 - t) - x_2(t - t_1))^2}{(t_2 - t)(t - t_1)(t_2 - t_1)}.$$

This can be seen by multiplying the previous line by $(t_2 - t)(t - t_1)(t_2 - t_1)$ and collecting the x^2 coefficients (of each side)

$$(t - t_1)(t_2 - t_1) + (t_2 - t)(t_2 - t_1) = (t_2 - t_1)^2,$$

then the x_2^2-coefficients

$$(t - t_1)(t_2 - t_1) - (t_2 - t)(t - t_1) = (t - t_1)^2$$

and the x_1^2-coefficients

$$(t_2 - t)(t_2 - t_1) - (t_2 - t)(t - t_1) = (t_2 - t)^2.$$

The agreement between the xx_2 coefficients, xx_1 coefficients, and x_1x_2 coefficients is also readily apparent.

The resulting PDF is rewritten as follows:

$$\frac{1}{\sqrt{2\pi \cdot \frac{(t-t_1)(t_2-t)}{t_2-t_1}}} \exp\left(-\frac{\left(x - \frac{x_1(t_2-t)+x_2(t-t_1)}{t_2-t_1}\right)^2}{2c\frac{(t_2-t)(t-t_1)}{t_2-t_1}}\right),$$

which is normal, with mean $\frac{x_1(t_2-t)+x_2(t-t_1)}{t_2-t_1}$ and variance $c \cdot \frac{(t_2-t)(t-t_1)}{t_2-t_1}$. Note the mean is just a linear interpolation of the x value at t connecting the (t_1, x_1) and (t_2, x_2) points and that the variance becomes zero at $t = t_1$ and $t = t_2$, as expected.

Example 10.3. Consider a Brownian motion with $d = -13 \frac{\text{cm}}{\text{h}}$ and $c = 124 \frac{\text{cm}^2}{\text{h}}$. At 8:04 a.m. the process was observed at 12.7 cm; at 10:26 a.m., it was at -4.7 cm. Find the probability that at 9:00 a.m. it had a value lower than 5 cm.

Solution.

$$> 1 - TailZ\left(\frac{5 - \dfrac{12.7 \cdot 86 - 4.7 \cdot 56}{142}}{\sqrt{124 \cdot \dfrac{86 \cdot 56}{142 \cdot 60}}}\right);$$

$$0.46013$$

$\square$

10.4 First-Passage Time

Our objective is to find the distribution of T, the time Brownian motion enters, for the first time, State a. At the same time, we want to find the condition that makes entering State a, sooner or later, certain.

To do this, we assume the process starts (at time 0) in State 0 and that T is the random time of visiting, for the first time, *either* the State $a > 0$ or the State $b < 0$ (this makes the issue easier to investigate; eventually, we will take the $b \to -\infty$ limit to answer the original question).

We know the moment-generation function (MGF) of X_t is given by

$$\mathbb{E}\left(\exp(uX_t)\right) = \exp\left(\tfrac{u^2 c \cdot t}{2} + u \cdot d \cdot t\right), \tag{10.4}$$

but it can also be expanded in the following manner:

$$\mathbb{E}\left(\exp(uX_t) \mid T \leq t\right) \cdot \Pr(T \leq t) + \mathbb{E}\left(\exp(uX_t) \mid T > t\right) \cdot \Pr(T > t)$$
$$= \mathbb{E}\left(\exp\left(u(X_t - X_T)\right) \cdot \exp\left(uX_T\right) \mid T \leq t\right) \cdot \Pr(T \leq t)$$
$$\quad + \mathbb{E}\left(\exp(uX_t) \mid T > t\right) \cdot \Pr(T > t)$$
$$= \mathbb{E}\left(\exp\left(\frac{u^2 c(t - T)}{2} + u \cdot d(t - T)\right) \cdot \exp\left(uX_T\right) \,\middle|\, T \leq t\right) \cdot \Pr(T \leq t)$$
$$\quad + \mathbb{E}\left(\exp(uX_t) \mid T > t\right) \cdot \Pr(T > t), \tag{10.5}$$

as $X_t - X_T$ and X_T are independent. Making this equal to the right-hand side of (10.4) and dividing the resulting equation by (10.4) yields

$$\mathbb{E}\left(\exp\left(-\frac{u^2 c \cdot T}{2} - u \cdot d \cdot T\right) \cdot \exp\left(uX_T\right) \,\middle|\, T \leq t\right) \cdot \Pr(T \leq t)$$
$$+ \exp\left(-\frac{u^2 c \cdot t}{2} - u \cdot d \cdot t\right) \cdot \mathbb{E}\left(\exp(uX_t) \mid T > t\right) \cdot \Pr(T > t) = 1. \tag{10.6}$$

We can now argue $\lim_{t \to \infty} \Pr(T > t) = 0$ and

$$\exp\left(-\frac{u^2 c \cdot t}{2} - u \cdot d \cdot t\right) \mathbb{E}\left(\exp(uX_t) \mid T > t\right)$$

remains bounded (whenever $\frac{u^2 c}{2} + u \cdot d \geq 0$, which is sufficient for our purposes), so that (in this limit) the last term of (10.6) disappears and (10.6) becomes (called, in this form, *Wald's identity*):

$$1 = \mathbb{E}\left(\exp\left(-\frac{u^2 c \cdot T}{2} - u \cdot d \cdot T\right) \cdot \exp\left(uX_T\right)\right)$$

$$= e^{u \cdot a} \cdot \mathbb{E}\left(\exp\left(-\frac{u^2 c \cdot T}{2} - u \cdot d \cdot T\right) \,\middle|\, X_T = a\right) \Pr(X_T = a)$$

$$+ e^{u \cdot b} \cdot \mathbb{E}\left(\exp\left(-\frac{u^2 c \cdot T}{2} - u \cdot d \cdot T\right) \,\middle|\, X_T = b\right) \Pr(X_T = b). \qquad (10.7)$$

There are two values of u for which $\exp\left(-\frac{u^2 c \cdot T}{2} - u \cdot d \cdot T\right)$ is identically equal to 1: 0 and $-\frac{2d}{c}$. Evaluating the left-hand side of (10.7) at $u = -\frac{2d}{c}$ yields

$$\exp\left(-\frac{2d}{c}a\right)\Pr(X_T = a) + \exp\left(-\frac{2d}{c}b\right)\Pr(X_T = b) = 1.$$

This, together with

$$\Pr(X_T = a) + \Pr(X_T = b) = 1,$$

enables us to solve for

$$\Pr(X_T = a) = \frac{\exp\left(-\frac{2d}{c}b\right) - 1}{\exp\left(-\frac{2d}{c}b\right) - \exp\left(-\frac{2d}{c}a\right)}$$

and

$$\Pr(X_T = b) = \frac{1 - \exp\left(-\frac{2d}{c}a\right)}{\exp\left(-\frac{2d}{c}b\right) - \exp\left(-\frac{2d}{c}a\right)}.$$

Now, by expanding the characteristic function of T,

$$\mathbb{E}\left(\exp(i\phi T)\right) = \mathbb{E}\left(\exp(i\phi T) \mid X_T = a\right) \Pr(X_T = a)$$
$$+ \mathbb{E}\left(\exp(i\phi T) \mid X_T = b\right) \Pr(X_T = b),$$

and solving

$$-\frac{u^2 c}{2} - u \cdot d = i\phi$$

for u, that is,

$$u_{1,2} = -\frac{d}{c} \pm \sqrt{\frac{d^2}{c^2} - \frac{2i\phi}{c}},$$

enables us to set up the following two equations, based on (10.7):

$$1 = e^{u_1 a} \cdot \mathbb{E}\left(\exp(i\phi T) \mid X_T = a\right) \Pr(X_T = a)$$
$$+ e^{u_1 b} \cdot \mathbb{E}\left(\exp(i\phi T) \mid X_T = b\right) \Pr(X_T = b),$$

$$1 = e^{u_2 a} \cdot \mathbb{E}\left(\exp(i\phi T) \mid X_T = a\right) \Pr(X_T = a)$$
$$+ e^{u_2 b} \cdot \mathbb{E}\left(\exp(i\phi T) \mid X_T = b\right) \Pr(X_T = b),$$

which can be solved for

$$\mathbb{E}\left(\exp(i\phi T) \mid X_T = a\right) \Pr(X_T = a) = \frac{e^{u_2 b} - e^{u_1 b}}{e^{u_1 a + u_2 b} - e^{u_1 b + u_2 a}}$$

and

$$\mathbb{E}\left(\exp(i\phi T) \mid X_T = b\right) \Pr(X_T = b) = \frac{e^{u_1 a} - e^{u_2 a}}{e^{u_1 a + u_2 b} - e^{u_1 b + u_2 a}}.$$

The sum of these two expressions is the characteristic function of T.

When $b \to -\infty$, the first of these expressions tends to

$$e^{-u_2 a} = \exp\left(\frac{a \cdot d}{c} - a\sqrt{\frac{d^2}{c^2} - \frac{2i\phi}{c}}\right) \tag{10.8}$$

and the second one tends to 0.

Let us see whether we can now convert the characteristic function (10.8) to the corresponding PDF.

Inverse Gaussian Distribution

Proposition 10.2. *The characteristic function (10.8) corresponds to the following PDF:*

$$f_T(t) = \frac{a}{\sqrt{2\pi \cdot c \cdot t^3}} \exp\left(-\frac{(d \cdot t - a)^2}{2c \cdot t}\right). \tag{10.9}$$

Proof. Using MAPLE:

> *assume* $(a > 0 \text{ and } d > 0 \text{ and } c > 0)$:

$$> CF := \int_0^\infty \frac{a \cdot \exp\left(\dfrac{(d \cdot t - a)^2}{2 \cdot c \cdot t} + I \cdot t \cdot u\right)}{\sqrt{2 \cdot \pi \cdot t^3}} \, dt;$$

$$CF := e^{-\dfrac{a \sim \left(d \sim + \sqrt{d \sim^2 - 2Iuc \sim}\right)}{c \sim}}$$

{ A "$\sim$" indicates the attached variable has assumptions associated with it.}   □

This enables us to answer any probability question about T.

Proposition 10.3. *The expected value of T is $\frac{a}{d}$, and its variance is $\frac{a \cdot c}{d^3}$.*

Proof. This is verified by differentiating (10.8).

$$> \mu := simplify\left(\left.\frac{\frac{\mathrm{d}}{\mathrm{d}u}CF}{I}\right|_{u=0}\right);$$

$$\mu := \frac{a \sim}{d \sim}$$

$$> var := simplify\left(-\left.\frac{\mathrm{d}^2}{\mathrm{d}u^2}CF\right|_{u=0} - \mu^2\right);$$

$$\mu := \frac{a \sim c \sim}{d \sim^2}$$

□

Proposition 10.4. *Equation (10.9) yields the following distribution func-tion:*

$$\Phi\left(\frac{d \cdot t - a}{\sqrt{c \cdot t}}\right) + \exp\left(\frac{2a \cdot d}{c}\right) \cdot \Phi\left(-\frac{d \cdot t + a}{\sqrt{c \cdot t}}\right),$$

where Φ is the distribution function of $\mathcal{N}(0,1)$.

Proof.

$$> \psi := z \to \frac{\int_{-\infty}^{z} \exp\left(-\frac{u^2}{2}\right) \mathrm{d}u}{\sqrt{2 \cdot \pi}}:$$

$$> simplify\left(\frac{\mathrm{d}}{\mathrm{d}t}\left(\psi\left(\frac{d \cdot t - a}{\sqrt{c \cdot t}}\right) + \psi\left(-\frac{d \cdot t + a}{\sqrt{c \cdot t}}\right) \cdot \exp\left(\frac{2 \cdot a \cdot d}{c}\right)\right)\right);$$

$$\frac{1}{2}\frac{\sqrt{2}e^{-\frac{1}{2}\frac{(-d \sim t + a \sim)^2}{c \sim t}} a \sim}{\sqrt{\pi} \sqrt{c \sim} t^{3/2}}$$

□

The limit of (10.9) when $d \to 0$ is

$$\frac{a}{\sqrt{2\pi \cdot c \cdot t^3}} \exp\left(-\frac{a^2}{2c \cdot t},\right)$$

in agreement with (10.3).

Example 10.4. Assuming $d = 1.12\frac{\mathrm{cm}}{\mathrm{sec}}$, $c = 3.8\frac{\mathrm{cm}^2}{\mathrm{s}}$ and $a = 11\mathrm{cm}$, display the PDF of the corresponding first-passage time.

What is the probability that, half a minute later, State a will not have been visited yet?

Solution.

$$> f := \frac{a \cdot \exp\left(-\dfrac{(d \cdot t - a)^2}{2 \cdot c \cdot t}\right)}{\sqrt{2 \cdot \pi \cdot c \cdot t^3}} :$$

$$> (d, c, a) := (1.12, 3.8, 11) :$$

$$> \int_{30.0}^{\infty} f \, dt;$$

$$0.007481$$

$$> plot\,(f, t = 0..40);$$

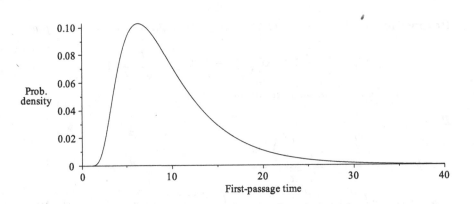

EXERCISES

Exercise 10.1. Consider one-dimensional Brownian motion with no drift, an absorbing barrier at zero, and $c = 3\,\frac{cm^2}{s}$, starting at $X(0) = 4$ cm. Calculate the probability of:

(a) The process getting absorbed within the first 15 s;
(b) $20\,cm > X(15\,s) > 10\,cm$.

Exercise 10.2. Similarly, assuming a Brownian motion with $c = 13.8\,\frac{cm^2}{h}$ and $d = 0$ (no absorbing barrier), find:

(a) $\Pr\,(X(3\,h) > -4\,cm \mid X(0) = 1\,cm)$;
(b) $\Pr\left(X(24h) > 15cm \cap \min_{0 < t < 1\text{day}} X(t) > 0 \mid X(0) = 3cm\right)$;

(c) $\Pr\left(\max_{0<t<1\text{day}} X(t) > 15\text{cm} \,\middle|\, X(0) = 0\right)$;

(d) The probability the process will return to its initial value (at least once) during the fourth hour (i.e., between 3 and 4 h later).

Exercise 10.3. Consider a Brownian motion with a drift of $-5.2\frac{\text{mm}}{\text{h}}$ and a diffusion coefficient of $7.3\frac{\text{mm}^2}{\text{h}}$. Evaluate:

(a) $\Pr(X(10:13) < 26\,\text{mm}\,|\,X(9:31) = 30\,\text{mm})$,
(b) $\Pr(X(10:13) < 26\,\text{mm}\,|\,X(9:31) = 30\,\text{mm} \cap X(10:42) = 27\,\text{mm})$.

Exercise 10.4. Consider a Brownian motion with drift equal to $-2.3\frac{\text{mm}}{\text{h}}$ and a diffusion coefficient of $12\frac{\text{mm}^2}{\text{h}}$. If the process is observed to have a value of 3.2 mm at 8:23 a.m., then a value of -1 mm at 10:41 a.m., and a value of -0.8 mm at 11:17 a.m., compute the probability that:

(a) It will have had a positive value at 9:30 a.m.;
(b) It will have had a positive value at 11:00 a.m.;
(c) It will have a positive value at 12:00.

Exercise 10.5. Consider a Brownian motion with drift equal to $-2.3\frac{\text{mm}}{\text{h}}$ and a diffusion coefficient of $12\frac{\text{mm}^2}{\text{h}}$. If the process is observed to have a value of 3.2 mm at 8:23 a.m., what is the probability that it will have had a negative value by 9:30 a.m.?

Exercise 10.6. Consider a Brownian motion with $d = -2.13\frac{\text{cm}}{\text{s}}$ and $c = 11.6\frac{\text{cm}^2}{\text{s}}$, $X(0) = 5.7$ cm, and an absorbing barrier at 0 cm. Find the expected value and standard deviation of the time till absorption.

CHAPTER 11
Autoregressive Models

We investigate processes having the following generalized Markovian property: to forecast their future, we need only the last k consecutive observations; a more distant past becomes irrelevant. We also discuss the issue of parameter estimation (unique to this chapter). The resulting theory can be applied to model random daily fluctuations (but not systematic or seasonal trends) of a stock market and similar situations.

11.1 BASICS

For the processes of this chapter time runs *discretely* (e.g., 0, 1, 2, 3, 4, ...), whereas $X_0, X_1, X_2, X_4, \ldots$ are random variables on a *continuous* scale. Processes of this type are called time series; here we investigate one special model, called an autoregressive model of such processes.

One of the most important concepts related to such processes is that of a SERIAL CORRELATION. The first-order serial correlation is defined as $\rho(X_i, X_{i+1})$, that is, the usual correlation coefficient between two consecutive values of a process. Similarly, the kth-order serial correlation is $\rho(X_i, X_{i+k})$. In general, these may all depend on the value of i, but when the process is *stationary* (which we assume, in this context, from now on), they become functions of the time lag k only (and are denoted by ρ_k). For completeness, we define the zero-order serial correlation as $\rho(X_i, X_i) = 1$.

A plot of the first handful of these serial correlation coefficients is called a CORRELOGRAM.

Let us look at some simple examples.

J. Vrbik and P. Vrbik, *Informal Introduction to Stochastic Processes with Maple*, Universitext, DOI 10.1007/978-1-4614-4057-4_11,
© Springer Science+Business Media, LLC 2013

WHITE NOISE

When all the X_i values are independent of each other and distributed normally with zero mean and some common standard deviation σ, the corresponding process is called WHITE NOISE. The sequence of serial correlations is very simple: except for $\rho_0 = 1$, they are all equal to zero.

MARKOV MODEL

We first start with a white-noise sequence $\varepsilon_1, \varepsilon_2, \varepsilon_3, \ldots$, and then generate the values of $X_0, X_1, X_2, \ldots$, by

$$X_{i+1} = \rho X_i + \varepsilon_{i+1}, \tag{11.1}$$

where X_0 is set arbitrarily (usually to 0; this makes the first several X_i nonstationary).

Note it is possible to express each X_i in terms of X_0 and the ε_j values (where $j \leq i$); thus,

$$X_1 = \rho X_0 + \varepsilon_1.$$
$$X_2 = \rho^2 X_0 + \rho \varepsilon_1 + \varepsilon_2.$$
$$X_3 = \rho^3 X_0 + \rho^2 \varepsilon_1 + \rho \varepsilon_2 + \varepsilon_3.$$
$$X_4 = \rho^4 X_0 + \rho^3 \varepsilon_1 + \rho^2 \varepsilon_2 + \rho \varepsilon_3 + \varepsilon_4.$$
$$\vdots$$

Also note $\text{Cov}(X_i, \varepsilon_j) = 0$ whenever $i < j$ (but not when $i \geq j$). Because each X_i is a linear combination of normally distributed ε, its distribution must be normal as well. When (and if) the process becomes stationary, all X must have the same distribution $\mathcal{N}(\mu, \Sigma)$, where μ and Σ are the corresponding mean and *standard deviation*, respectively. To find these (in terms of X_0, σ, and ρ, the only parameters of the process), we first take the expected value of (11.1), assuming i is large enough for the process to have become stationary. We get $\mu = \rho \cdot \mu + 0$, which implies $\mu = 0$. Similarly, if we *square* each side of (11.1) before taking the expected value, we get

$$\Sigma^2 = \rho^2 \Sigma^2 + \sigma^2$$

(because ε_{i+1} and X_i are independent), implying

$$\Sigma = \sqrt{\frac{\sigma^2}{1 - \rho^2}}.$$

Note the last formula becomes infinite at $\rho = 1$ and $\rho = -1$. This tells us that the process becomes asymptotically stationary, as i increases, only when $-1 < \rho < 1$, and goes wild otherwise.

It is interesting to note if we make X_0 itself random, having a normal distribution with a mean of zero and a standard deviation equal to Σ, then the process is stationary from the very beginning.

Example 11.1. Generate and plot 50 consecutive values based on a stationary Markov model with $\rho = 0.87$ and $\sigma = 1$.

Solution.

```
> ρ := 0.87 :
> σ := 1 :
> X := [0$50] : {A list of 50 zeroes}
```

$$> X_1 := Sample\left(Normal\left(0, \frac{\sigma}{\sqrt{1 - \rho^2}}\right), 1\right)_1 :$$

```
> for i from 1 to 49 do
      X_{i+1} := ρ · X_i + Sample (Normal (0, 1) , 1_1) :
   end do:
> listplot(X);
```

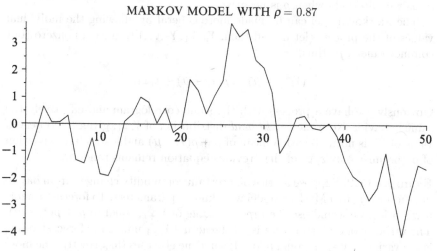

MARKOV MODEL WITH $\rho = 0.87$

As already stated, we are interested only in the *stationary* behavior of the processes in this chapter. Under those circumstances, we establish the value of the first-order serial correlation of the Markov model by multiplying each side of (11.1) by X_i and taking the expected value. This yields

$$\text{Cov}(X_{i+1}, X_i) = \rho \cdot \text{Cov}(X_i, X_i) + \text{Cov}(\varepsilon_{i+1}, X_i) = \rho \text{Var}(X_i)$$

because X_i and ε_{i+1} are independent. Dividing by the stationary variance Σ^2 yields

$$\rho_1 = \frac{\rho \mathrm{Var}(X_i)}{\Sigma^2} = \rho.$$

Similarly, to get the second-order serial correlation, we multiply both sides of (11.1) by X_{i-1} and take the expected value,

$$\mathrm{Cov}(X_{i+1}, X_{i-1}) = \rho \cdot \mathrm{Cov}(X_i, X_{i-1}) + \mathrm{Cov}(\varepsilon_{i+1}, X_{i-1}) = \rho \cdot \mathrm{Cov}(X_i, X_{i-1}),$$

which yields (after dividing by Σ^2)

$$\rho_2 = \frac{\rho \cdot \rho \cdot \Sigma^2}{\Sigma^2} = \rho^2.$$

In this manner, one can continue to get the following general result:

$$\rho_n = \rho^n.$$

The corresponding correlogram is thus a simple geometric progression (the signs will alternate when ρ is negative).

The Markov model can be made more general by allowing the individual values of the process (let us call them $Y_0, Y_1, Y_2, \ldots$) to have a nonzero (but common) mean μ; thus,

$$(Y_{i+1} - \mu) = \rho(Y_i - \mu) + \varepsilon_{i+1}.$$

Obviously, we have a process with the same correlogram and effectively the same behavior. Note the *conditional* distribution of Y_{i+1}, given the observed value of Y_i, is normal, with a mean of $\mu + \rho(Y_i - \mu)$ and a standard deviation of σ (because only ε_{i+1} of the previous equation remains random).

Example 11.2. Suppose a price of a certain commodity changes, from day to day, according to a Markov model with known parameters. To forecast tomorrow's price, we would use the *expected value* of Y_{i+1}, equal to $\mu + \rho(Y_i - \mu)$. The variance of our error, that is, of the actual Y_{i+1} minus the forecast value, is the variance of ε_{i+1}, equal to σ^2. If someone else uses (incorrectly) the more conservative white-noise model, that prediction would always be μ. Note the variance of the error would be $\mathrm{Var}(Y_{i+1} - \mu) = \frac{\sigma^2}{1-\rho^2}$. Using $\rho = 0.9$, this is 5.26 times larger than the variance of the error of *our* prediction. $\Diamond$

Example 11.3 (Extension of Example 11.2).   With the help of two-dimensional integration, we now compute the *probability* of our forecast being closer to the actual value (when it is observed) than the conservative, white-noise prediction, that is,

$$\Pr\left(|Y_{i+1} - \mu| > |Y_{i+1} - \mu - \rho(Y_i - \mu)|\right) = \Pr\left(|\rho(Y_i - \mu) + \varepsilon_{i+1}| > |\varepsilon_{i+1}|\right),$$

where $\rho(Y_i - \mu)$ and ε_{i+1} are independent and normal and with a mean of zero (each) and a variance of $\frac{\rho^2 \sigma^2}{1-\rho^2}$ and σ^2, respectively.

By introducing $Z_1 \equiv \frac{\varepsilon_{i+1}}{\sigma}$ and $Z_2 \equiv \frac{\rho(Y_i - \mu)}{\rho\sigma}\sqrt{1-\rho^2}$ (two independent *standardized normal* random variables), the same probability can be written as

$$\Pr\left(\left| \frac{\rho\sigma Z_2}{\sqrt{1-\rho^2}} + \sigma Z_1 \right| > |\sigma Z_1| \right) = \Pr\left(\left| \frac{\rho Z_2}{\sqrt{1-\rho^2}} + Z_1 \right| > |Z_1| \right),$$

where (assuming $\rho > 0$) the last inequality has two distinct solutions:

$$Z_1 > -\frac{\rho Z_2}{2\sqrt{1-\rho^2}} \text{ for } Z_1 > 0$$

and

$$Z_1 < -\frac{\rho Z_2}{2\sqrt{1-\rho^2}} \text{ for } Z_1 < 0$$

(to see this, one must first try four possibilities, taking each $\frac{\rho Z_2}{\sqrt{1-\rho^2}} + Z_1$ and Z_1 to be either positive or negative). The final answer is

$$\frac{1}{2\pi} \iint\limits_{R_1} \exp\left(-\frac{z_1^2 + z_2^2}{2} \right) \, dz_1 \, dz_2 + \frac{1}{2\pi} \iint\limits_{R_2} \exp\left(-\frac{z_1^2 + z_2^2}{2} \right) \, dz_1 \, dz_2$$

$$= \frac{2}{2\pi} \iint\limits_{R_1} \exp\left(-\frac{z_1^2 + z_2^2}{2} \right) \, dz_1 \, dz_2$$

$$= \frac{1}{\pi} \int\limits_0^\infty r \exp\left(-\frac{r^2}{2} \right) \, dr \int\limits_0^{\frac{\pi}{2} + \arctan \theta_0} d\theta$$

$$= \frac{1}{2} + \frac{1}{\pi} \arctan\left(\frac{\rho}{2\sqrt{1-\rho^2}} \right),$$

where $\theta_0 \equiv \frac{\rho}{2\sqrt{1-\rho^2}}$.

A similar analysis with $\rho < 0$ would yield

$$\frac{1}{2} + \frac{1}{\pi} \arctan\left(\frac{|\rho|}{2\sqrt{1-\rho^2}} \right),$$

which is correct regardless of the sign of ρ. For $\rho = 0.9$, this results in 75.51%, a noticeably better "batting average." $\diamondsuit$

11.2 YULE MODEL

The previous (Markov) model generated the next value based on the current observation only. To incorporate the possibility of following a trend (i.e., stock prices are on the rise), we have to go a step further, using

$$X_i = \alpha_1 X_{i-1} + \alpha_2 X_{i-2} + \varepsilon_i. \tag{11.2}$$

Note X_i now represents the *next* (tomorrow's) value, and X_{i-1}, X_{i-2}, and so on are the *current* and *past* (today's and yesterday's) observations.

Example 11.4. Generate a realization of this process using

1. $\alpha_1 = 0.2$, $\alpha_2 = 0.7$,
2. $\alpha_1 = 1.8$, $\alpha_2 = -0.9$,
3. $\alpha_1 = -1.8$, $\alpha_2 = -0.9$,

where $X_1 = 0$ and $X_2 = 0$.

Solution.

```
> x := Vector(300, 0) :
> ε := Sample (Normal(0, 1), 300) :
> for i from 3 to 300 do
      x_i := 0.2 · x_{i-1} + 0.7 · x_{i-2} + 3 · ε_i;
  end do:
> listplot(x[50.. − 1]);
```

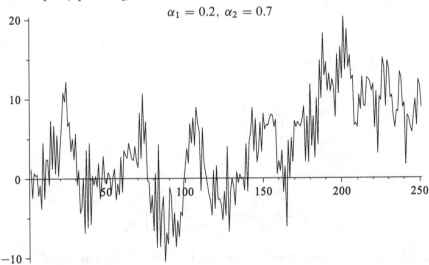

$$\alpha_1 = 0.2, \ \alpha_2 = 0.7$$

```
> for i from 3 to 300 do
      x_i := 1.8 · x_{i-1} − 0.9 · x_{i-2} + 3 · ε_i;
```

end do:
> $listplot(x[50.. - 1]);$

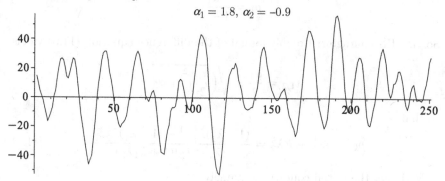

$$\alpha_1 = 1.8, \; \alpha_2 = -0.9$$

> **for** i **from** 3 **to** 300 **do**
  $x_i := -1.8 \cdot x_{i-1} - 0.9 \cdot x_{i-2} + 3 \cdot \varepsilon_i;$
  end do:
> $listplot(x[50.. - 1]);$

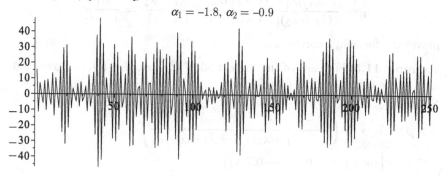

$$\alpha_1 = -1.8, \; \alpha_2 = -0.9$$

One can observe that, depending on the values of the α parameters, we obtain totally different types of behavior. $\square$

Assuming the process has reached its stationary state of affairs (serial correlation no longer depends on i, and all variances are identical), we get, upon multiplying the previous formula by X_{i-k} and taking the expected value (recall the mean value of each X_i is zero and that X_{i-k} and ε_i are independent) the following result:

$$\rho_k = \alpha_1 \rho_{k-1} + \alpha_2 \sigma_{k-2} \tag{11.3}$$

when $k = 2, 3, \ldots$, and

$$\rho_1 = \alpha_1 \rho_0 + \alpha_2 \rho_1$$

when $k = 1$. Since $\rho_0 = 1$, based on the last equation, we get

$$\rho_1 = \frac{\alpha_1'}{1 - \alpha_2}.$$

Solving the characteristic polynomial of the difference equation (11.3) yields

$$\lambda_{1,2} = \frac{\alpha_1}{2} \pm \sqrt{\left(\frac{\alpha_1}{2}\right)^2 + \alpha_2},$$

so that

$$\rho_k = A\lambda_1^k + B\lambda_2^k = \frac{(1 - \lambda_2^2)\lambda_1^{k+1} - (1 - \lambda_1^2)\lambda_2^{k+1}}{(\lambda_1 - \lambda_2)(1 + \lambda_1\lambda_2)}. \tag{11.4}$$

Verifying the initial conditions, namely,

$$\rho_0 = \frac{(1 - \lambda_2^2)\lambda_1 - (1 - \lambda_1^2)\lambda_2}{(\lambda_1 - \lambda_2)(1 + \lambda_1\lambda_2)} = 1$$

and

$$\rho_1 = \frac{(1 - \lambda_2^2)\lambda_1^2 - (1 - \lambda_1^2)\lambda_2^2}{(\lambda_1 - \lambda_2)(1 + \lambda_1\lambda_2)} = \frac{\lambda_1 + \lambda_2}{1 + \lambda_1\lambda_2} = \frac{\alpha_1}{1 - \alpha_2},$$

proves the formula's correctness.

Example 11.5. For each of the three models of Example 11.4, compute and display the corresponding correlogram.

Solution.

> $\rho := \lambda \rightarrow \mathrm{Re}\left(\dfrac{\left(1 - \lambda_2^2\right) \cdot \lambda_1^{k+1} - \left(1 - \lambda_1^2\right) \cdot \lambda_2^{k+1}}{(\lambda_1 - \lambda_2) \cdot (1 + \lambda_1 \cdot \lambda_2)}\right)$:

> $\lambda := \left[solve\left(x^2 = 0.2 \cdot x + 0.7, x\right)\right];$

$$\lambda := [0.9426, -0.7426]$$

> $listplot\left(\left[seq\left([k, \rho(\lambda)]\right), k = 0..50\right]\right);$

$$\alpha_1 = 0.2, \ \alpha_2 = 0.7$$

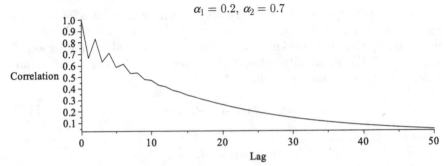

$> \lambda := \left[solve\left(x^2 = 1.8 \cdot x - 0.9, x \right) \right];$

$$\lambda := [0.9000 + 0.3000\,\mathbf{I}, 0.9000 - 0.3000\,\mathbf{I}]$$

$> listplot\left([seq\left([k, \rho(\lambda)] \right), k = 0..70] \right);$

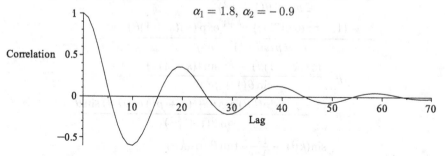

$> \lambda := \left[solve\left(x^2 = -1.8 \cdot x - 0.9, x \right) \right];$

$$\lambda := [-0.9000 + 0.3000\,\mathbf{I}, -0.9000 - 0.3000\,\mathbf{I}]$$

$> listplot\left([seq\left([k, \rho(\lambda)] \right), k = 0..70] \right);$

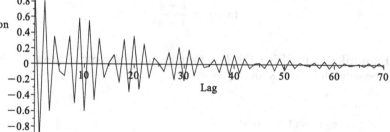

When $\lambda_1 = \lambda_2$, (11.4) needs to be modified by setting $\lambda_1 = \lambda + \varepsilon$ and $\lambda_2 = \lambda$ and taking the $\varepsilon \to 0$ limit. This yields

$$\begin{aligned}
\rho_k &= \lim_{\varepsilon \to 0} \frac{(1 - \lambda^2)(\lambda + \varepsilon)^{k+1} - \left(1 - (\lambda + \varepsilon)^2 \right) \lambda^{k+1}}{\varepsilon(1 + \lambda^2)} \\
&= \lim_{\varepsilon \to 0} \frac{(1 - \lambda^2)(k + 1)\lambda^k \varepsilon + 2\lambda\varepsilon\lambda^{k+1}}{\varepsilon(1 + \lambda^2)} \\
&= \left(1 + \frac{1 - \lambda^2}{1 + \lambda^2} \cdot k \right) \lambda^k.
\end{aligned}$$

When the two λ roots are complex conjugate, say $\lambda_{1,2} = p \cdot \exp(\pm i\theta)$, the expression for ρ_k can be converted into an explicitly *real* form by

$$
\frac{\left(1 - p^2 \exp(-2i\theta)\right) p^{k+1} \exp\left(i(k+1)\theta\right)}{2pi\sin\theta(1+p^2)}
$$
$$
- \frac{-\left(1 - p^2 \exp(2i\theta)\right) p^{k+1} \exp\left(-i(k+1)\theta\right)}{2pi\sin\theta(1+p^2)}
$$
$$
= p^k \frac{\sin\left((k+1)\theta\right) - p^2 \sin\left((k-1)\theta\right)}{\sin\theta(1+p^2)}
$$
$$
= p^k \frac{(1-p^2)\sin(k\theta)\cos\theta + (1+p^2)\cos(k\theta)\sin\theta}{\sin\theta(1+p^2)}
$$
$$
= p^k \frac{\sin(k\theta) + \frac{1+p^2}{1-p^2}\tan\theta\cos(k\theta)}{\frac{1+p^2}{1-p^2}\tan\theta}
$$
$$
= p^k \frac{\sin(k\theta + \psi)}{\sin\psi},
$$

where

$$
\tan\psi \equiv \frac{1+p^2}{1-p^2}\tan\theta.
$$

Example 11.6. Using the last formula, plot the correlogram of the Yule process with $\alpha_1 = 1.8$ and $x_2 = -0.9$.

Solution.

```
> (α₁, α₂) := (1.8, −0.9) :
> (θ, p) := solve (z² − α₁ · z − α₂, z) :
> θ := argumet (θ); p := abs(p); ψ := arctan ( (1+p²)/(1−p²) · tan (θ) );
```

$$\theta := 0.3218 \qquad p := 0.9487 \qquad \psi := 1.4142$$

```
> listplot ([ seq ([k, (pᵏ · sin (k · θ + ψ))/sin (ψ) ], k = 0..50) ]);
```

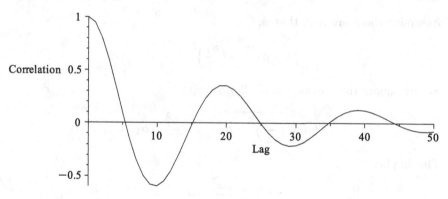

{Getting the same answer as in Example 11.5, using a different formula.}

$\square$

Squaring each side of (11.2) and taking the expected value we now obtain

$$\mathrm{Var}(X_i) = \alpha_1^2 \mathrm{Var}(X_{i-1}) + \alpha_2^2 \mathrm{Var}(X_{i-2}) + 2\alpha_1\alpha_2 \mathrm{Cov}(X_{i-1}, X_{i-2}) + \sigma^2,$$

which implies

$$\Sigma^2 \left(1 - \alpha_1^2 - \alpha_2^2 - \frac{2\alpha_1^2\alpha_2}{1 - \alpha_2} \right) = \sigma^2, \qquad (11.5)$$

as all three variances are identical and equal to Σ^2, and

$$\mathrm{Cov}(X_{i-1}, X_{i-2}) = \rho_1 \Sigma^2 = \frac{\alpha_1}{1 - \alpha_2} \Sigma^2.$$

Finally, (11.5) implies

$$\Sigma^2 = \frac{1 - \alpha_2}{(1 + \alpha_2)(1 - \alpha_1 - \alpha_2)(1 + \alpha_1 - \alpha_2)} \sigma^2. \qquad (11.6)$$

An alternate (but equivalent) expression for Σ^2 can be obtained by multiplying (11.2) by X_i and taking the expected value, to get

$$\Sigma^2 = \alpha_1 \Sigma^2 \rho_1 + \alpha_2 \Sigma^2 \rho_2 + \sigma^2,$$

and solving for Σ^2. The reader should verify $\mathbb{E}(X_i, \varepsilon_i) = \sigma^2$.

STABILITY ANALYSIS

The Yule model is stable (i.e., to say: asymptotically stationary) when both λ are, in absolute value, smaller than 1:

$$\left| \frac{\alpha_1}{2} \pm \sqrt{\left(\frac{\alpha_1}{2}\right)^2 + \alpha_2} \right| < 1.$$

Assuming the λ are real, that is,

$$\alpha_2 \geq -\left(\frac{\alpha_1}{2}\right)^2,$$

we can square the previous inequality, getting

$$\pm\alpha_1 \sqrt{\left(\frac{\alpha_1}{2}\right)^2 + \alpha_2} < 1 - \frac{\alpha_1^2}{2} - \alpha_2.$$

This implies

$$0 < 1 - \frac{\alpha_1^2}{2} - \alpha_2$$

and, upon another squaring,

$$\alpha_1^2 \left(\left(\frac{\alpha_1}{2}\right)^2 + \alpha_2\right) < \left(1 - \frac{\alpha_1^2}{2} - \alpha_2\right)^2,$$

which is equivalent to

$$0 < 1 - 2\alpha_2 - \alpha_1^2 + \alpha_2^2 = (1 - \alpha_2 - \alpha_1)(1 - \alpha_2 + \alpha_1).$$

This further implies $\alpha_2 < 1 - \alpha_1$ and $\alpha_2 < 1 + \alpha_1$ since the opposite ($>$ and $>$) would contradict a previous equation. Together with $\alpha_2 \geq -\left(\frac{\alpha_1}{2}\right)^2$, this yields one region of possibilities.

Assuming the λ are complex conjugates, that is,

$$\alpha_2 < -\left(\frac{\alpha_1}{2}\right)^2,$$

we can square the left-hand side of our condition by multiplying each side by its own complex conjugate:

$$\left(\frac{\alpha_1}{2} + i\sqrt{-\left(\frac{\alpha_1}{2}\right)^2 - \alpha_2}\right)\left(\frac{\alpha_1}{2} - i\sqrt{-\left(\frac{\alpha_1}{2}\right)^2 - \alpha_2}\right) = -\alpha_2 < 1,$$

implying

$$\alpha_2 > -1.$$

Together with $\alpha_2 < -\left(\frac{\alpha_1}{2}\right)^2$, this yields the second part of the stable region. Combining the two parts together results in the following triangle:

$$\alpha_2 < 1 - \alpha_1 \quad \text{and} \quad \alpha_2 < 1 + \alpha_1 \quad \text{and} \quad \alpha_2 > -1.$$

Note that this agrees with the region inside the $\Sigma^2 = \infty$ boundary; see (11.6).

PARTIAL SERIAL CORRELATION

Proposition 11.1. *For the Markov model, the partial correlation (Appendix 11.A) between X_i and X_{i-2}, given the value of X_{i-1}, is*

$$\frac{\rho_2 - \rho^2}{\sqrt{1-\rho^2}\sqrt{1-\rho^2}} = 0$$

since $\rho_2 = \rho^2$ (this is, of course, the expected answer). On the other hand, the same partial correlation for the Yule model yields (since $\rho_2 = \frac{\alpha_1^2}{1-\alpha_2} + \alpha_2$)

$$\frac{\frac{\alpha_1^2}{1-\alpha_2} + \alpha_2 - \left(\frac{\alpha_1}{1-\alpha_2}\right)^2}{\sqrt{1-\left(\frac{\alpha_1}{1-\alpha_2}\right)^2}\sqrt{1-\left(\frac{\alpha_1}{1-\alpha_2}\right)^2}} = \frac{\alpha_2(1-\alpha_2)^2 - \alpha_1^2\alpha_2}{(1-\alpha_2)^2 - \alpha_1^2} = \alpha_2.$$

This provides a natural interpretation of the α_2 coefficient.

For the same Yule model, we get a zero partial correlation between X_i and X_{i-3} given the values of X_{i-1} and X_{i-2}, as expected.

Proof.

$$\rho_{14|3} = \frac{\frac{\alpha_1^3 + \alpha_1\alpha_2}{1-\alpha_2} + \alpha_1\alpha_2 - \left(\frac{\alpha_1^2}{1-\alpha_2} + \alpha_2\right)\frac{\alpha_1}{1-\alpha_2}}{\sqrt{1-\left(\frac{\alpha_1^2}{1-\alpha_2} + \alpha_2\right)^2}\sqrt{1-\left(\frac{\alpha_1}{1-\alpha_2}\right)^2}} = \frac{\alpha_1\alpha_2}{\sqrt{1+\alpha_1^2 - \alpha_2^2}},$$

$$\rho_{12|3} = \frac{\frac{\alpha_1}{1-\alpha_2} - \left(\frac{\alpha_1^2}{1-\alpha_2} + \alpha_2\right)\frac{\alpha_1}{1-\alpha_2}}{\sqrt{1-\left(\frac{\alpha_1^2}{1-\alpha_2} + \alpha_2\right)^2}\sqrt{1-\left(\frac{\alpha_1}{1-\alpha_2}\right)^2}} = \frac{\alpha_1}{\sqrt{1+\alpha_1^2 - \alpha_2^2}},$$

$$\rho_{24|3} = \rho_{13,2} = \frac{\frac{\alpha_1^2}{1-\alpha_2} + \alpha_2 - \left(\frac{\alpha_1}{1-\alpha_2}\right)^2}{\sqrt{1-\left(\frac{\alpha_1}{1-\alpha_2}\right)^2}\sqrt{1-\left(\frac{\alpha_1}{1-\alpha_2}\right)^2}} = \alpha_2,$$

implying

$$\rho_{14|23} = \frac{\rho_{14|3} - \rho_{12,3} \cdot \rho_{24|3}}{\sqrt{1-\rho_{12|3}^2}\sqrt{1-\rho_{24|3}^2}} = 0. \qquad \square$$

11.3 GENERAL AUTOREGRESSIVE MODEL

We can go beyond the Yule model (which usually increases the model's predictive power, at the cost of making it more complicated) by using

$$X_i = \alpha_1 X_{i-1} + \alpha_2 X_{i-2} + \alpha_3 X_{i-3} + \varepsilon_i$$

or, if this is still not enough,

$$X_i = \alpha_1 X_{i-1} + \alpha_2 X_{i-2} + \alpha_3 X_{i-3} + \alpha_4 X_{i-4} + \varepsilon_i$$

etc. In general, any such model is called AUTOREGRESSIVE. Finding the corresponding formulas for ρ_k and $\mathrm{Var}(X_i)$ becomes increasingly more difficult, so we will first deal with particular cases only.

Example 11.7. Assuming a time series is generated via

$$X_i = 0.3X_{i-1} + 0.1X_{i-2} - 0.2X_{i-3} + \varepsilon_i,$$

where the ε_i are independent, $\mathcal{N}(0, \sqrt{15})$ type random variables (white noise), and assuming that we are observing a stationary part of the sequence, we can find the serial correlation coefficients from

$$\rho_k = 0.3\rho_{k-1} + 0.1\rho_{k-2} - 0.2\rho_{k-3}, \tag{11.7}$$

where $k = 3, 4, 5, \ldots$, and

$$\rho_2 = 0.3\rho_1 + 0.1\rho_0 - 0.2\rho_1,$$
$$\rho_1 = 0.3\rho_0 + 0.1\rho_1 - 0.2\rho_2.$$

The last two equations imply

```
> solns := solve({ρ2 = 0.3 · ρ1 + 0.1 · ρ0 − 0.2 · ρ1,
      ρ1 = 0.3 · ρ0 + 0.1 · ρ1 − 0.2 · ρ2, ρ0 = 1}, {ρ0, ρ1, ρ2});
```

$$solns := \{\rho_0 = 1.0, \rho_1 = 0.3043, \rho_2 = 0.1304\}$$

```
> assign(solns); {This will assign the preceding values of ρ so we can use
them.}
{Equation (11.7) enables us to continue,}
> for i from 3 to 10 do
      ρi := 0.3 · ρi−1 + 0.1 · ρi−2 − 0.2 · ρi−3;
   end do:
> listplot(convert(ρ,list));
```

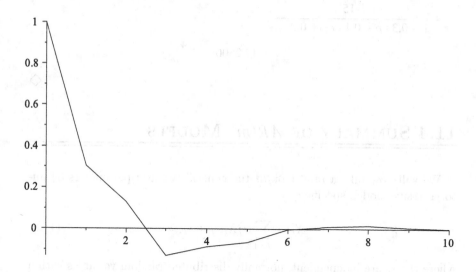

To obtain an expression for any ρ_k, we must first solve the corresponding cubic polynomial.

> $\lambda := solve\left(\lambda^3 = 0.3 \cdot \lambda^2 + 0.1 \cdot \lambda - 0.2, \lambda\right) :$

> $\varrho := k \rightarrow a \cdot \lambda_1^k + b \cdot \lambda_2^k + c \cdot \lambda_3^k :$

> $solns := solve\left(\{\varrho(0) = 1, \varrho(1) = \rho_1, \varrho(2) = \rho_2\}, \{a, b, c\}\right);$

$$solns := \{a = 0.3951 - 0.09800\mathrm{I}, \; b = 0.2099, \; c = 0.3951 + 0.09800\mathrm{I}\}$$

> $assign(solns):$

This yields the following general formula.

> $\varrho(k);$

$$(0.3951 - 0.09800\mathrm{I})(0.4241 + 0.4302\mathrm{I})^k + 0.2099\,(-0.5481)^k$$
$$+ (0.3951 + 0.09800\mathrm{I})(0.4241 - 0.4302\mathrm{I})^k$$

And, to verify it against the previous results,

> $seq\left(\mathrm{Re}\left(\varrho(k)\right), k = 3..10\right);$

$$-0.1304, \; -0.0870, \; -0.06522, \; -0.002174, \; 0.01022,$$

$$0.01589, \; 0.006224, \; 0.001413$$

The variance of the X follows from (to be justified in the next section)

$$> \frac{15}{1 - 0.3 \cdot \rho - 0.1 \cdot \rho_2 + 0.2 \cdot \rho_3};$$

$$17.2500$$

11.4 SUMMARY OF $AR(m)$ MODELS

We will now take a brief look at the general (with m parameters α) autoregressive model, specified by

$$X_i = \sum_{j=1}^{m} \alpha_j X_{i-j} + \varepsilon_i, \tag{11.8}$$

where the ε_i are independent, normally distributed random variables with a mean of zero and standard deviation of σ.

The previous equation implies (assuming a stationary situation has been reached) that the X_i are also normally distributed, with a mean of zero and a variance of Σ^2 (same for all X_i), and that the correlation coefficient between X_i and X_{i+k} (denoted ρ_k) is independent of i.

Proposition 11.2. *Given $\rho_0 = 1$, the remaining ρ can be computed from*

$$\rho_k = \sum_{j=1}^{m} \alpha_j \rho_{|k-j|}$$

where $k = 1, 2, 3, \ldots$.

Proof. Multiply (11.8) by X_{i-k}, take the expected value, and divide by Σ^2.
□

The first $m - 1$ of these equations can be solved for $\rho_1, \rho_2, \ldots, \rho_{m-1}$; the remaining equations then provide a recursive formula to compute $\rho_m, \rho_{m+1}, \ldots$.

Proposition 11.3. *The common variance, Σ^2, of the X_i is*

$$\Sigma^2 = \frac{\sigma^2}{1 - \sum_{j=1}^{m} \alpha_j \rho_j}.$$

Proof. Multiply (11.8) by X_i, take the expected value of each side, and solve for Σ^2.
□

The variance–covariance matrix $\mathbb{V}$ of n consecutive X_i consists of the following elements:

$$\mathbb{V}_{ij} = \Sigma^2 \cdot \rho_{|i-j|}.$$

Proposition 11.4. *When $n \geq 2m$, $\mathbb{V}$ has a surprisingly simple band-matrix inverse $\mathbb{A} = \mathbb{V}^{-1}$, with elements given by*

$$\mathbb{A}_{ij} = \frac{1}{\sigma^2} \sum_{\ell=0}^{\text{Min}(i-1,j-1,n-i,n-j)} \alpha_\ell \alpha_{\ell+|i-j|}, \qquad (11.9)$$

with the understanding that $\alpha_0 = -1$ and $\alpha_\ell = 0$ when $\ell > m$.

Proof. Firstly, the corresponding probability density function (PDF) can be written as a product of the PDF of the first m of these and of

$$(2\pi)^{(n-m)/2} \exp\left(-\frac{\sum_{i=m+1}^{n}(x_i - \sum_{j=1}^{m}\alpha_j x_{i-j})^2}{2\sigma^2}\right).$$

Secondly, the resulting $\mathbb{A}$ must be both symmetric and slant (i.e., /) symmetric since $\mathbb{V}$ has both of these properties.   □

The corresponding *determinant* cannot be expressed in terms of a general formula, but it can be easily evaluated for any m (amazingly, aside from the trivial scaling factor of σ^{-2n}, it has the *same* value for all $n \geq m$). It can always be factorized in the following manner:

$$D \equiv \det(\mathbb{A}) = \sigma^{-2n} \left(\sum_{j=0}^{m} \alpha_j\right)\left(\sum_{j=0}^{m}(-1)^j \alpha_j\right) S_m^2$$

(using the same $\alpha_0 = -1$ convention), where

m	S_m
1	1
2	$1 + \alpha_2$
3	$1 + \alpha_2 + \alpha_1\alpha_3 - \alpha_3^2$
4	$1 + \alpha_2 + \alpha_1\alpha_3 - \alpha_3^2 +$ $\alpha_4(1 + \alpha_1^2 + 2\alpha_2 - \alpha_1\alpha_3)$ $-\alpha_4^2(1 - \alpha_2) - \alpha_4^3$
⋮	⋮

or, alternatively,

$$D = \sigma^{-2n} \prod_{i,j=1}^{m} (1 - \lambda_i \lambda_j),$$

where the λ are the m solutions to

$$\lambda^m = \sum_{j=1}^{m} \alpha_j \lambda^{m-j}.$$

Note the denominator of Σ^2 is

$$\left(\sum_{j=0}^{m} \alpha_j \right) \left(\sum_{j=0}^{m} (-1)^j \alpha_j \right) S_m,$$

which leads to the following simple conditions to ensure the process's stability:

$$\sum_{j=0}^{m} \alpha_j < 0,$$

$$\sum_{j=0}^{m} (-1)^j \alpha_j < 0,$$

$$S_m > 0.$$

The last condition is actually a bit more involved – it requires S_m to be positive everywhere on the line connecting the *origin* and the $(\alpha_1, \alpha_2, \ldots, \alpha_m)$ point in the corresponding m-dimensional space.

The multivariate PDF of n consecutive X_n is thus

$$f(\mathbf{x}) = \frac{\sqrt{D} \exp\left(-\frac{\mathbf{x}^T \mathbb{A} \mathbf{x}}{2} \right)}{(2\pi)^{\frac{n}{2}}}. \tag{11.10}$$

Example 11.8. Find and display the three-dimensional region (in the α_1, α_2, and α_3 space) inside which the AR(3) model is stable.

Solution.

```
> srf₁ := −1 + α₁ + α₂ + α₃ :
> srf₂ := −1 − α₁ + α₂ − α₃ :
> srf₃ := 1 + α₂ + α₁ · α₂ − α₃² :
> solve (srf₁ = srf₂, α₃);
```
$$-\alpha_1$$

```
> solve (srf₁ = srf₃, α₃);
```
$$1, -2 + \alpha_1$$

```
> solve (srf₁ = srf₂, α₃);
```
$$-1, 2 + \alpha_1$$

```
> plt₁ := plot3d (1 − α₁ − α₃, α₃ = −1..1, α₁ = −α₃..α₃ + 2) :
```

> $plt_2 := plot3d\,(1 + \alpha_1 + \alpha_3,\ \alpha_3 = -1..1,\ \alpha_1 = -2 + \alpha_3..-\alpha_3)$:
> $plt_3 := plot3d\,(\alpha_3^2 - 1 - \alpha_1 \cdot \alpha_3,\ \alpha_3 = -1..1,\ \alpha_1 = -2 + \alpha_3..\alpha_3 + 2)$:
> $display\,(plt_1,\ plt_2,\ plt_3,\ axes = boxed)$;

$AR(3)$ STABILITY REGION

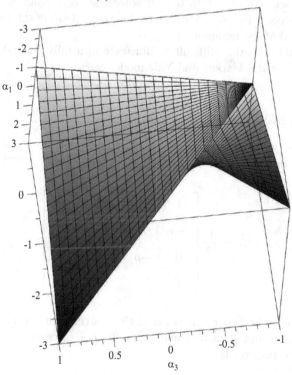

11.5 PARAMETER ESTIMATION

In practice, it is important to be able to estimate the value of all parameters of an $AR(K)$, based on a sequence of n consecutive observations. The best way of doing this is by maximizing the so-called LIKELIHOOD FUNCTION [the expression on the right-hand side of (11.10), where x is replaced by the vector of observations – plain numbers – and the parameters α_j, μ, and σ are now considered variable]. Or equivalently, by maximizing its logarithm, namely,

$$n \ln{(2\pi)} + 2n \ln{(\sigma)} - \ln\left(-\sum_{j=0}^{K} \alpha_j\right) - \ln\left(\sum_{j=0}^{K}(-1)^{i+1}\alpha_j\right) - 2\ln(S_K)$$

$$+ (\mathbf{x} - \boldsymbol{\mu})^{\mathrm{T}} \cdot \mathbb{A} \cdot (\mathbf{x} - \boldsymbol{\mu}). \tag{11.11}$$

Note, for future convenience, we have multiplied the logarithm by -2 (implying we will be minimizing, instead of maximizing); also, both $\sum_{j=0}^{K} \alpha_j$ and $\sum_{j=0}^{K}(-1)^i \alpha_j$ have been multiplied by -1 to make them positive.

All that is required now is to differentiate (11.11) with respect to each parameter, set the answer to 0, and solve the corresponding set of NOR-MAL EQUATIONS. The solution yields the MAXIMUM-LIKELIHOOD ESTIMA-TORS (MLEs) of the parameters.

This would be rather difficult to illustrate in a fully general form, so we do this only for the Markov and Yule model cases.

MARKOV MODEL

This time, we have only one α, which is customary to denote by ρ. This implies (11.9) simplifies to

$$
\mathbb{A} = \frac{1}{\sigma^2}
\begin{bmatrix}
1 & -\rho & 0 & \cdots \\
-\rho & 1+\rho^2 & -\rho & \ddots \\
0 & -\rho & 1+\rho^2 & \ddots \\
\vdots & \ddots & \ddots & \ddots
\end{bmatrix}
$$

displaying only the upper left corner of the matrix (which is tridiagonal and has both main and slant-diagonal symmetry). Expression (11.11), further divided by n, thus reads

$$
\ln(2\pi) + 2\ln(\sigma) - \frac{\ln(1-\rho^2)}{n} - \frac{\rho^2 (Z_1^2 + Z_n^2)}{n} + (1+\rho^2)\overline{Z^2} - 2\rho\overline{ZZ_{-1}}, \tag{11.12}
$$

where

$$
Z_i \equiv \frac{X_i - \mu}{\sigma},
$$

$$
\overline{Z^2} \equiv \frac{1}{n} \sum_{i=1}^{n} Z_i^2,
$$

$$
\overline{ZZ_{-j}} \equiv \frac{1}{n} \sum_{i=1+j}^{n} Z_i Z_{i-j},
$$

and $X_1, X_2, \ldots, X_n$ are the n consecutive observations.

Maximum-Likelihood Estimators

Differentiating (11.11) with respect to each μ, σ, and ρ leads to the following three equations:

$$\widehat{\mu} = \frac{\overline{X} + \frac{\widehat{\rho}(X_1 + X_n)}{n(1 - \widehat{\rho})}}{1 + \frac{2\widehat{\rho}}{n(1 - \widehat{\rho})}},$$

$$\widehat{\sigma}^2 = \left(1 + \widehat{\rho}^2\right)\overline{(X - \widehat{\mu})^2} - 2\widehat{\rho}\overline{(X - \widehat{\mu})(X - \widehat{\mu})_{-1}} - \widehat{\rho}^2 \frac{(X_1 - \widehat{\mu})^2 + (X_n - \widehat{\mu})^2}{n},$$

$$\widehat{\rho} = \frac{\overline{(X - \widehat{\mu})(X - \widehat{\mu})_{-1}} - \frac{\widehat{\rho}\widehat{\sigma}^2}{n(1 - \widehat{\rho}^2)}}{\overline{(X - \widehat{\mu})^2} - \frac{(X_1 - \widehat{\mu})^2 + (X_n - \widehat{\mu})^2}{n}},$$

where the hat symbol, "$\widehat{}$," implies $\widehat{\mu}$, $\widehat{\sigma}$, and $\widehat{\rho}$ are no longer the exact parameters but their *estimators* (each is a *random variable* with its own distribution).

The most expedient way to solve these equations is to use MAPLE. (We require the X from Example 11.1; furthermore, since we use a model with $\mu = 0$, we need to estimate σ and ρ only.)

```
> n := nops (X) :

> Z := evalm ( X/σ ) :

> LF := 2 · n · ln(σ) − ln (1 − ρ²) − ρ² · (Z₁² + Zₙ²) + (1 + ρ²) · Σⁿᵢ₌₁ Zᵢ²
        −2 · p · Σⁿᵢ₌₂ (Zᵢ · Zᵢ₋₁) :

> fsolve ( { ∂/∂σ LF, ∂/∂ρ LF }, {σ = 0..∞, ρ = −1..1} );
```

$$\{\rho = 0.8395, \ \sigma = 0.8898\}$$

The program returns the corresponding MLEs of σ and ρ.

Yule Model

To find the MLEs μ, σ, α_1, and α_2, we now need to minimize

$$\ln(2\pi) + 2\ln(\sigma) - \frac{\ln(1 - \alpha_1 - \alpha_2) + \ln(1 + \alpha_1 - \alpha_2) + 2\ln(1 + \alpha_2)}{n}$$

$$- \frac{\left(\alpha_1^2 + \alpha_2^2\right)\left(Z_1^2 + Z_n^2\right) + \alpha_2^2\left(Z_2^2 + Z_{n-1}^2\right) + 2\alpha_1\alpha_2\left(Z_1 Z_2 + Z_{n-1} Z_n\right)}{n}$$

$$+ \left(1 + \alpha_1^2 + \alpha_2^2\right)\overline{Z^2} - 2\alpha_1\left(1 - \alpha_2\right)\overline{ZZ_{-1}} - 2\alpha_2\overline{ZZ_{-2}} \tag{11.13}$$

since now

$$
\mathbb{A} = \frac{1}{\sigma^2}
\begin{bmatrix}
1 & -\alpha_1 & -\alpha_2 & 0 & \cdots \\
\alpha_1 & 1 + \alpha_1^2 & -\alpha_1\,(1 - \alpha_2) & -\alpha_2 & \ddots \\
\alpha_2 & -\alpha_1\,(1 - \alpha_2) & 1 + \alpha_1^2 + \alpha_2^2 & -\alpha_1\,(1 - \alpha_2) & \ddots \\
0 & -\alpha_2 & -\alpha_1\,(1 - \alpha_2) & 1 + \alpha_1^2 + \alpha_2^2 & \ddots \\
\vdots & \ddots & \ddots & \ddots & \ddots
\end{bmatrix}
$$

and

$$
D = \frac{(1 - \alpha_1 - \alpha_2)\,(1 + \alpha_1 - \alpha_2)\,(1 + \alpha_2)^2}{\sigma^{2n}}.
$$

Again, assuming $\mu = 0$, we find the σ, α_1, and α_2 estimators by

```
> x := [0$100] : {100 zeros.}
> ε := Sample (Normal(0, 1), 100) :
> for i from 3 to 100 do do
    xᵢ := 0.2 · xᵢ₋₁ + 0.7 · xᵢ₋₂ + 3 · εᵢ;
  end do:
```

$> x := 0.2 \cdot x_{i-1} + 0.7 \cdot x_{i-2} + 3 \cdot \varepsilon_i;$

> $x := x\,[51.. - 1]$: {Let us consider only the last 50 equilibrated values.}

> $n := nops\,(x) :\ Z := evalm\left(\dfrac{x}{\sigma}\right) :$

> $unassign(`i`):$

> $LF := 2 \cdot n \cdot \ln(\sigma) - \ln\,(1 - \alpha_1 - \alpha_2) - \ln\,(1 + \alpha_1 - \alpha_2) - 2 \cdot \ln\,(1 + \alpha_2)$
$\qquad - \left(\alpha_1^2 + \alpha_2^2\right) \cdot \left(Z_1^2 + Z_n^2\right) - \alpha_2^2 \cdot \left(Z_2^2 + Z_{n-1}^2\right)$

$\qquad\qquad - 2 \cdot \alpha_1 \cdot \alpha_2 \cdot \left(Z_1 \cdot Z_2 + Z_{n-1} \cdot Z_n\right) + \left(1 + \alpha_1^2 + \alpha_2^2\right) \cdot \displaystyle\sum_{i=1}^{n} Z_i^2$

$\qquad\qquad - 2 \cdot \alpha_1 \cdot (1 - \alpha_2) \cdot \displaystyle\sum_{i=2}^{n} Z_i \cdot Z_{i-1} - 2 \cdot \alpha_2 \cdot \displaystyle\sum_{i=3}^{n} Z_i \cdot Z_{i-2} :$

> $solns := fsolve\left(\left\{\dfrac{\partial}{\partial\sigma} LF,\ \dfrac{\partial}{\partial\alpha_1} LF,\ \dfrac{\partial}{\partial\alpha_2} LF\right\},\right.$

$\qquad\qquad \left. \{\sigma = 0..\infty,\ \alpha_1 = -2..2,\ \alpha_2 = -1..1\}\right);$

> $assign(solns):\ \sigma = 3.2068,\ \alpha_1 = 0.2021,\ \alpha_2 = -0.6634$

> $\alpha_2 < 1 - \alpha_1;\ \alpha_2 < 1 + \alpha_1$

$$0.6634 < 0.7979$$

$$0.6634 < 1.2021$$

{Here we verify the solutions are inside the stability region (as luck would have it, they are) – if they were outside, a new solution would have to be found using the *avoid* = {*solns*} option.}

11.A NORMAL DISTRIBUTION AND PARTIAL CORRELATION

UNIVARIATE NORMAL DISTRIBUTION

In general, a normal distribution has two parameters, μ and σ (mean and standard deviation). A special case is a STANDARDIZED normal distribution, with a mean of 0 and standard deviation of 1. A general X can be converted into a standardized Z by

$$Z = \frac{X - \mu}{\sigma}$$

and reverse

$$X = \sigma Z + \mu.$$

It is usually a lot easier to deal with Z and then convert the results back into X.

In this context recall that when $X \in \mathcal{N}(\mu, \sigma)$,

$$aX + b \in \mathcal{N}(a\mu + b, |a|\sigma), \tag{11.14}$$

where a and b are constants.

The PDFs of Z and X are

$$f_Z(z) = \frac{\exp\left(-\frac{z^2}{2}\right)}{\sqrt{2\pi}},$$

$$f_X(x) = \frac{\exp\left(-\frac{(x-\mu)^2}{2\sigma^2}\right)}{\sqrt{2\pi}\,\sigma}$$

respectively.

Similarly, the corresponding moment-generating functions (MGFs) are

$$M_Z(t) = \exp\left(\frac{t^2}{2}\right) \quad \text{and}$$

$$M_X(t) = e^{\mu t} \cdot M_Z(\sigma t) = \exp\left(\frac{\sigma^2 t^2}{2} + \mu t\right).$$

BIVARIATE NORMAL DISTRIBUTION

Again, we consider two versions of this distribution, the general (X and Y) and standardized (Z_1 and Z_2) distributions. The general distribution is defined by five parameters (the individual means and variances, plus the correlation coefficient ρ); the standardized version has only one parameter, namely ρ.

The corresponding joint (bivariate) PDFs are

$$f_{zz}(z_1, z_2) = \frac{\exp\left(-\dfrac{z_1^2 + z_2^2 - 2\rho z_1 z_2}{2(1-\rho^2)}\right)}{2\pi\sqrt{1-\rho^2}}$$

and

$$f_{xy}(x, y) = \frac{\exp\left(-\dfrac{(\frac{x-\mu_1}{\sigma_1})^2 + (\frac{y-\mu_2}{\sigma_2})^2 - 2\rho(\frac{x-\mu_1}{\sigma_1})(\frac{y-\mu_2}{\sigma_2})}{2(1-\rho^2)}\right)}{2\pi\sigma_1\sigma_2\sqrt{1-\rho^2}}$$

for the standardized and general cases, respectively.

Similarly, the joint MGFs are

$$M_{zz}(t_1, t_2) = \exp\left(\frac{t_1^2 + t_2^2 + 2\rho t_1 t_2}{2}\right)$$

and

$$M_{xy}(t_1, t_2) = e^{\mu_1 t_1 + \mu_2 t_2} \cdot M_{zz}(\sigma_1 t_1, \sigma_2 t_2)$$
$$= \exp\left(\frac{\sigma_1^2 t_1^2 + \sigma_2^2 t_2^2 + 2\rho\sigma_1\sigma_2 t_1 t_2}{2} + \mu_1 t_1 + \mu_2 t_2\right).$$

We should remember a joint MGF enables us to find joint moments of the distribution by

$$\mathbb{E}\left(X^n Y^m\right) = \left.\frac{\partial^{(n+m)} M_{xy}(t_1, t_2)}{\partial t_1^n \partial t_2^m}\right|_{t_1 = t_2 = 0}.$$

Also, we can easily find the MGF of a MARGINAL distribution of X by setting $t_2 = 0$. This tells us immediately that $x \in \mathcal{N}(\mu_1, \sigma_1)$ and that both Z_1 and Z_2 are standardized normal.

CONDITIONAL DISTRIBUTION

The CONDITIONAL DISTRIBUTION of $Z_1 \mid Z_2 = \underline{z}_2$ (an underline implies that $\underline{z}_2$ is no longer a variable but is assumed to have a specific, observed value).

Finding the corresponding (univariate) PDF is done by

$$\frac{f_{zz}(z_1, \underline{z}_2)}{f_z(\underline{z}_2)} = \frac{\dfrac{\exp\left(-\dfrac{z_1^2 + \underline{z}_2^2 - 2\rho z_1 \underline{z}_2}{2(1 - \rho^2)}\right)}{2\pi\sqrt{1 - \rho^2}}}{} \div \frac{\exp\left(-\dfrac{\underline{z}_2^2}{2}\right)}{\sqrt{2\pi}}$$

$$= \frac{\exp\left(-\dfrac{(z_1 - \rho\underline{z}_2)^2}{2(1 - \rho^2)}\right)}{\sqrt{2\pi}\sqrt{1 - \rho^2}}.$$

The result can be identified as $\mathcal{N}(\rho\underline{z}_2, \sqrt{1 - \rho^2})$, that is, normal, with a mean of $\rho\underline{z}_2$ and standard deviation equal to $\sqrt{1 - \rho^2}$ (smaller than what it was marginally, i.e., before we observed Z_2).

We now utilize this result to find the conditional distribution of X given that Y has been observed to have a value of $\underline{y}$ (instead of using a similar, direct approach, requiring rather messy algebra).

We already know the conditional distribution of

$$\frac{X - \mu_1}{\sigma_1} \quad \text{given that} \quad \frac{Y - \mu_2}{\sigma_2} = \frac{\underline{y} - \mu_2}{\sigma_2}$$

is

$$\mathcal{N}\left(\rho\frac{\underline{y} - \mu_2}{\sigma_2}, \sqrt{1 - \rho^2}\right).$$

Consequently, we know the conditional distribution of

$$\frac{X - \mu_1}{\sigma_1} \quad \text{given that} \quad Y = \underline{y},$$

which is the same thing. Now, by a linear-transformation argument, we find the conditional distribution of $X \mid Y = \underline{y}$ is

$$\mathcal{N}\left(\mu_1 + \sigma_1\rho\frac{\underline{y} - \mu_2}{\sigma_2}, \sigma_1\sqrt{1 - \rho^2}\right).$$

MULTIVARIATE NORMAL DISTRIBUTION

Consider N independent, standardized, normally distributed random variables. Their joint PDF is the product of the individual PDFs

$$f(z_1, z_2, \ldots, z_N) = (2\pi)^{-N/2} \cdot \exp\left(-\frac{\displaystyle\sum_{i=1}^{N} z_i^2}{2}\right)$$

$$= (2\pi)^{-N/2} \cdot \exp\left(-\frac{\underline{z}^T \underline{z}}{2},\right)$$

and the corresponding MGF is, likewise,

$$\exp\left(\frac{\sum\limits_{i=1}^{N} t_i^2}{2}\right) = \exp\left(\frac{t^T t}{2}\right).$$

The linear transformation

$$\mathbf{X} = \mathbb{B}\mathbf{Z} + \boldsymbol{\mu},$$

where $\mathbb{B}$ is an arbitrary (regular) $N \times N$ matrix, defines a new set of N random variables having a *general* MULTIVARIATE normal distribution. The corresponding PDF is

$$\frac{|\det(\mathbb{B}^{-1})|}{\sqrt{(2\pi)^N}} \exp\left(-\frac{(\mathbf{x} - \boldsymbol{\mu})^T (\mathbb{B}^{-1})^T \mathbb{B}^{-1} (\mathbf{x} - \boldsymbol{\mu})}{2}\right)$$

$$= \frac{1}{\sqrt{(2\pi)^N \det(\mathbb{V})}} \exp\left(-\frac{(\mathbf{x} - \boldsymbol{\mu})^T \mathbb{V}^{-1} (\mathbf{x} - \boldsymbol{\mu})}{2}\right),$$

where $\mathbb{V} \equiv \mathbb{B}\mathbb{B}^T$ is the corresponding VARIANCE–COVARIANCE MATRIX of $\mathbf{X}$ (it must be symmetric and POSITIVE DEFINITE). Since $\mathbf{Z} = \mathbb{B}^{-1}(\mathbf{X} - \boldsymbol{\mu})$, this further implies

$$(\mathbf{X} - \boldsymbol{\mu})^T (\mathbb{B}^{-1})^T \mathbb{B}^{-1} (\mathbf{X} - \boldsymbol{\mu}) = (\mathbf{X} - \boldsymbol{\mu})^T (\mathbb{B}\mathbb{B}^T)^{-1} (\mathbf{X} - \boldsymbol{\mu})$$

$$= (\mathbf{X} - \boldsymbol{\mu})^T \mathbb{V}^{-1} (\mathbf{X} - \boldsymbol{\mu})$$

must have a χ_N^2 distribution.

The corresponding multivariate MGF is

$$M_{\mathbf{X}}(t) = \mathbb{E}\left(\exp\left(t^T (\mathbb{B}\mathbf{Z} + \boldsymbol{\mu})\right)\right)$$

$$= \exp\left(t^T \boldsymbol{\mu}\right) \cdot \exp\left(\frac{t^T \mathbb{B}\mathbb{B}^T t}{2}\right)$$

$$= \exp\left(t^T \boldsymbol{\mu}\right) \cdot \exp\left(\frac{t^T \mathbb{V}t}{2}\right).$$

This shows each marginal distribution remains normal, without a change in the respective $\boldsymbol{\mu}$ and $\mathbb{V}$ elements.

Note there are many different $\mathbb{B}$ that result in the same $\mathbb{V}$. Generating a set of normally distributed random variables having a given variance–covariance matrix $\mathbb{V}$ requires us to find one such $\mathbb{B}$. The easiest way to construct $\mathbb{B}$ is to make it LOWER TRIANGULAR.

Example 11.9. Generate a random vector of five values from a normal distribution with the variance–covariance matrix equal to

$$\begin{bmatrix} 7 & 3 & 0 & 5 & 5 \\ 3 & 10 & 6 & -1 & -3 \\ 0 & 6 & 8 & -1 & -5 \\ 5 & -1 & -1 & 7 & 2 \\ 5 & -3 & -5 & 2 & 13 \end{bmatrix}$$

and the five means given by $\langle -1, 2, 0, 3, -2 \rangle$.

Solution.

$$> \mathbb{V} := \begin{bmatrix} 7 & 3 & 0 & 5 & 5 \\ 3 & 10 & 6 & -1 & -3 \\ 0 & 6 & 8 & -1 & -5 \\ 5 & -1 & -1 & 7 & 2 \\ 5 & -3 & -5 & 2 & 13 \end{bmatrix} :$$

$> n := 5 :$

$> \mathbb{B} := Matrix(n, n) :$

$> $ **for** i **from** 1 **to** n **do**

$$\mathbb{B}_{i,i} := \sqrt{\mathbb{V}_{i,i} - \sum_{k=1}^{i-1} {}^{\backprime}\mathbb{B}{}^{\prime 2}_{i,k}};$$

{Note the quotes around $\mathbb{B}$ are necessary for a technical reason.}

for j **from** $i + 1$ **to** n

$$\mathbb{B}_{j,i} := \frac{\mathbb{V}_{j,i} - \sum_{k=1}^{i-1} {}^{\backprime}\mathbb{B}{}^{\prime}_{i,k} \cdot {}^{\backprime}\mathbb{B}{}^{\prime}_{j,k}}{\mathbb{B}_{i,i}}$$

    end do

  end do:

$> \mathbb{B}.Transpose(\mathbb{B});$

{just to verify}

$$\begin{bmatrix} 7 & 3 & 0 & 5 & 5 \\ 3 & 10 & 6 & -1 & -3 \\ 0 & 6 & 8 & -1 & -5 \\ 5 & -1 & -1 & 7 & 2 \\ 5 & -3 & -5 & 2 & 13 \end{bmatrix}$$

> $evalm(evalf(\mathbb{B}).convert(Sample(Normal(0,1),5), vector)$
  $+[-1, 2, 0, 3, -2]);$

$$\begin{bmatrix} -3.8374 & -0.0437 & 0.6195 & 1.0471 & -3.8561 \end{bmatrix}$$

$\square$

FINDING MLEs OF μ AND V

Proposition 11.5. *Let $a_{k,\ell}$ be the kth-row, ℓth-column element of a square matrix $\mathbb{A}$; then*

$$\frac{\partial \ln(\det(\mathbb{A}))}{\partial a_{k,\ell}} = \left(\mathbb{A}^{-1}\right)_{\ell,k},$$

$$\frac{\partial (\mathbb{A}^{-1})_{i,j}}{\partial a_{k,\ell}} = -\left(\mathbb{A}^{-1}\right)_{i,k}\left(\mathbb{A}^{-1}\right)_{\ell,j}.$$

Proof. The determinant of a matrix can be expanded with respect to the kth row; thus,

$$\det(\mathbb{A}) = \sum_i (-1)^{k+i} a_{k,i} M_{k,i},$$

where $M_{k,i}$ is the corresponding MINOR (determinant of $\mathbb{A}$ with the kth row and ith column removed).

Utilizing

$$\frac{\partial a_{k,i}}{\partial a_{k,\ell}} = \delta_{i,\ell},$$

where $\delta_{i,\ell}$ (Kronecker's delta) is 1 when $i = \ell$ and 0 otherwise, we get

$$\frac{\partial \det(\mathbb{A})}{\partial a_{k,\ell}} = \sum_i (-1)^{k+i} \delta_{\ell,i} M_{k,i} = (-1)^{k+\ell} M_{k,\ell} = \left(\mathbb{A}^{-1}\right)_{\ell,k} \cdot \det(\mathbb{A})$$

and, thus,

$$\frac{\partial \ln(\det(\mathbb{A}))}{\partial a_{k,\ell}} = \frac{\partial \det(\mathbb{A})}{\partial a_{k,\ell}} \div \det(\mathbb{A}) = \left(\mathbb{A}^{-1}\right)_{\ell,k}.$$

To prove the second formula, we start with

$$\frac{\partial \mathbb{A}_{i,j}}{\partial a_{k,\ell}} = \delta_{i,k}\delta_{j,\ell} \tag{11.15}$$

and

$$\sum_m \mathbb{A}_{i,m}(\mathbb{A}^{-1})_{m,j} = \delta_{i,j}.$$

Differentiating the last identity with respect to $a_{k,\ell}$ yields

$$\sum_m \frac{\partial \mathbb{A}_{i,m}}{\partial a_{k,\ell}}(\mathbb{A}^{-1})_{m,j} + \sum_m \mathbb{A}_{i,m}\frac{\partial(\mathbb{A}^{-1})_{m,j}}{\partial a_{k,\ell}} = 0.$$

With the help of (11.15) we get

$$\sum_m \delta_{i,k}\delta_{m,\ell}(\mathbb{A}^{-1})_{m,j} + \sum_m \mathbb{A}_{i,m}\frac{\partial(\mathbb{A}^{-1})_{m,j}}{\partial a_{k,\ell}} = 0$$

or, equivalently,

$$\sum_m \mathbb{A}_{i,m}\frac{\partial(\mathbb{A}^{-1})_{m,j}}{\partial a_{k,\ell}} = -\delta_{i,k}(\mathbb{A}^{-1})_{\ell,j}.$$

Premultiplying by $(\mathbb{A}^{-1})_{n,i}$ and summing over i results in

$$\sum_{m,i}(\mathbb{A}^{-1})_{n,i}\mathbb{A}_{i,m}\frac{\partial(\mathbb{A}^{-1})_{m,j}}{\partial a_{k,\ell}} = -\sum_i(\mathbb{A}^{-1})_{n,i}\delta_{i,k}(\mathbb{A}^{-1})_{\ell,j},$$

from which follow

$$\sum_m \delta_{n,m}\frac{\partial(\mathbb{A}^{-1})_{m,j}}{\partial a_{k,\ell}} = -(\mathbb{A}^{-1})_{n,k}(\mathbb{A}^{-1})_{\ell,j}$$

and

$$\frac{\partial(\mathbb{A}^{-1})_{n,j}}{\partial a_{k,\ell}} = -(\mathbb{A}^{-1})_{n,k}(\mathbb{A}^{-1})_{\ell,j}.$$

$\square$

Taking ln of the likelihood function of a sample of n from a multivariate normal distribution, we get

$$-\frac{1}{2}\sum_{i=1}^n(\mathbf{X}_i - \boldsymbol{\mu})^{\mathrm{T}}\mathbb{V}^{-1}(\mathbf{X}_i - \boldsymbol{\mu}) - \frac{N \cdot n}{2}\ln(2\pi) - \frac{n}{2}\ln(\det(\mathbb{V})). \tag{11.16}$$

Differentiating

$$-\frac{1}{2}\sum_{i,j,k}(\mathbf{X}_i - \mu)_j(\mathbb{V}^{-1})_{j,k}(\mathbf{X}_i - \mu)_k$$

with respect to μ_ℓ yields

$$\frac{1}{2}\sum_{i,j,k}\delta_{j,\ell}(\mathbb{V}^{-1})_{j,k}(\mathbf{X}_i - \mu)_k + \frac{1}{2}\sum_{i,j,k}(\mathbf{X}_i - \mu)_j(\mathbb{V}^{-1})_{j,k}\delta_{k,\ell}$$

$$= \frac{1}{2}\sum_{i,k}(\mathbb{V}^{-1})_{\ell,k}(\mathbf{X}_i - \mu)_k + \frac{1}{2}\sum_{i,j}(\mathbf{X}_i - \mu)_j(\mathbb{V}^{-1})_{j,\ell}$$

$$= \sum_{i,k}(\mathbb{V}^{-1})_{\ell,k}(\mathbf{X}_i - \mu)_k,$$

which is the ℓth component of

$$\mathbb{V}^{-1}\sum_i(\mathbf{X}_i - \mu).$$

Making these equal to zero and solving for μ results in the expected answer of

$$\widehat{\mu} = \overline{\mathbf{X}}.$$

Differentiating (11.16) with respect to $v_{\ell,m}$ yields

$$\frac{1}{2}\sum_{i,j,k}(\mathbf{X}_i - \mu)_j(\mathbb{V}^{-1})_{j,\ell}(\mathbb{V}^{-1})_{m,k}(\mathbf{X}_i - \mu)_k - \frac{n}{2}(\mathbb{V}^{-1})_{m,\ell}$$

when $\ell = m$, double the previous expression when $\ell \neq m$. In either case, the corresponding normal equation reads

$$\sum_{j,k}(\mathbb{V}^{-1})_{m,k}\mathbb{S}_{k,j}(\mathbb{V}^{-1})_{j,\ell} = n \cdot (\mathbb{V}^{-1})_{m,\ell}$$

or, equivalently,

$$\mathbb{V}^{-1}\mathbb{S}\mathbb{V}^{-1} = n \cdot \mathbb{V}^{-1},$$

where

$$\mathbb{S}_{k,j} \equiv \sum_{i=1}^n(\mathbf{X}_i - \mu)_k(\mathbf{X}_i - \mu)_j.$$

Solving for $\mathbb{V}_{k,j}$ (and substituting $\overline{\mathbf{X}}$ for μ) yields

$$\widehat{\mathbb{V}}_{k,j} = \frac{\sum_{i=1}^n(\mathbf{X}_i - \overline{\mathbf{X}})_k(\mathbf{X}_i - \overline{\mathbf{X}})_j}{n}.$$

PARTIAL CORRELATION COEFFICIENT

A variance–covariance matrix can be converted into the following correlation matrix:

$$\mathbb{C}_{ij} \equiv \frac{\mathbb{V}_{ij}}{\sqrt{\mathbb{V}_{ii} \cdot \mathbb{V}_{jj}}}.$$

The main-diagonal elements of $\mathbb{C}$ are all equal to 1 (the correlation coefficient of X_i with itself).

Suppose we have three normally distributed random variables with a given variance–covariance matrix. The conditional distribution of X_2 and X_3, given that $X_1 = \underline{x}_1$, has a correlation coefficient independent of the value of $\underline{x}_1$. It is called the PARTIAL CORRELATION COEFFICIENT and is denoted by $\rho_{23|1}$. Let us find its value in terms of ordinary correlation coefficients.

All correlation coefficients are independent of scaling. We can thus choose the three X to be standardized (but *not* independent), having the following three-dimensional PDF:

$$\frac{1}{\sqrt{(2\pi)^3 \det(\mathbb{C})}} \cdot \exp\left(-\frac{\mathbf{x}^T \mathbb{C}^{-1} \mathbf{x}}{2}\right),$$

where

$$\mathbb{C} = \begin{bmatrix} 1 & \rho_{12} & \rho_{13} \\ \rho_{12} & 1 & \rho_{23} \\ \rho_{13} & \rho_{23} & 1 \end{bmatrix}.$$

Since the marginal PDF of X_1 is

$$\frac{1}{\sqrt{2\pi}} \cdot \exp\left(-\frac{x_1^2}{2}\right),$$

the conditional PDF of X_2 and X_3 given $X_i = \underline{x}_1$ is

$$\frac{1}{\sqrt{(2\pi)^2 \det(\mathbb{C})}} \cdot \exp\left(-\frac{\mathbf{x}^T \mathbb{C}^{-1} \mathbf{x} - \underline{x}_1^2}{2}\right).$$

The information about the five parameters of the corresponding bivariate distribution is in

$$\mathbf{x}^T \mathbb{C}^{-1} \mathbf{x} - \underline{x}_1^2 = \frac{z_1^2 + z_2^2 - 2\dfrac{\rho_{23} - \rho_{12}\rho_{13}}{\sqrt{1 - \rho_{12}^2}\sqrt{1 - \rho_{13}^2}} \cdot z_1 \cdot z_2}{1 - \left(\dfrac{\rho_{23} - \rho_{12}\rho_{13}}{\sqrt{1 - \rho_{12}^2}\sqrt{1 - \rho_{13}^2}}\right)^2},$$

where

$$z_1 = \frac{x_2 - \rho_{12}\underline{x}_1}{\sqrt{1 - \rho_{12}^2}},$$

$$z_2 = \frac{x_3 - \rho_{13}\underline{x}_1}{\sqrt{1 - \rho_{13}^2}},$$

which, in terms of the two conditional means and standard deviations, agrees with what we know already. The new information is our partial correlation coefficient

$$\rho_{23|1} = \frac{\rho_{23} - \rho_{12} \cdot \rho_{13}}{\sqrt{1 - \rho_{12}^2}\sqrt{1 - \rho_{13}^2}}$$

or

$$\rho_{ij|k} = \frac{\rho_{ij} - \rho_{ik} \cdot \rho_{jk}}{\sqrt{1 - \rho_{ik}^2}\sqrt{1 - \rho_{jk}^2}}$$

in general.

To get the conditional mean, standard deviation, and correlation coefficient given that more than one X has been observed, one can iterate the last formula, together with the conditional-mean/variance formulas, in the following manner:

$$\mu_{i|K\ell} = \mu_{i\ |K} + \sigma_{i\ |K} \cdot \rho_{i\ell\ |K} \cdot \frac{x_\ell - \mu_{\ell\ |K}}{\sigma_{\ell\ |K}},$$

$$\sigma_{i|K\ell} = \sigma_{i\ |K} \cdot \sqrt{1 - \rho_{i\ell\ |K}^2},$$

$$\rho_{ij|K\ell} = \frac{\rho_{ij|K} - \rho_{i\ell|K} \cdot \rho_{j\ell|K}}{\sqrt{1 - \rho_{i\ell|K}^2} \cdot \sqrt{1 - \rho_{j\ell|K}^2}},$$

etc., where K now represents any number of indices (corresponding to the already observed X).

A more direct way to find these is presented in the following section.

GENERAL CONDITIONAL DISTRIBUTION

Let N variables be partitioned into two subsets $\mathbf{X}_{(1)}$ and $\mathbf{X}_{(2)}$ with corresponding means $\boldsymbol{\mu}_{(1)}$ and $\boldsymbol{\mu}_{(2)}$ and the variance-covariance matrix of

$$\begin{bmatrix} \mathbf{V}_{11} & \mathbf{V}_{12} \\ \mathbf{V}_{21} & \mathbf{V}_{22} \end{bmatrix}. \tag{11.17}$$

Proposition 11.6. *The inverse of (11.17) is*

$$A = \begin{bmatrix} (V_{11} - V_{12}V_{22}^{-1}V_{21})^{-1} & -(V_{11} - V_{12}V_{22}^{-1}V_{21})^{-1}V_{12}V_{22}^{-1} \\ -(V_{22} - V_{21}V_{11}^{-1}V_{12})^{-1}V_{21}V_{11}^{-1} & (V_{22} - V_{21}V_{11}^{-1}V_{12})^{-1} \end{bmatrix}.$$

Proof. It is readily shown that $AV = \mathbb{I}$.

$VA = \mathbb{I}$ is an immediate consequence of Proposition 11.6, yielding four interesting identities.

Proposition 11.7. *The conditional PDF of $\mathbf{X}_{(1)}$ given $\mathbf{X}_{(2)} = \underline{\mathbf{x}}_{(2)}$ is*

$$\frac{1}{\sqrt{(2\pi)^N \det(V)}} \exp\left(-\frac{(\mathbf{x} - \boldsymbol{\mu})^T V^{-1} (\mathbf{x} - \boldsymbol{\mu})}{2} \right)$$

$$\div \frac{1}{\sqrt{(2\pi)^{N_2} \det(V_{22})}} \exp\left(-\frac{(\underline{\mathbf{x}}_{(2)} - \boldsymbol{\mu}_{(2)})^T V_{22}^{-1} (\underline{\mathbf{x}}_{(2)} - \boldsymbol{\mu}_{(2)})}{2} \right)$$

i.e., still normal. To get the resulting (conditional) variance–covariance matrix, all we need to do is invert the corresponding (i.e., the 1st-1st) block of A, getting

$$V_{(1|2)} \equiv V_{11} - V_{12}V_{22}^{-1}V_{21}.$$

Similarly, the conditional mean (say $\boldsymbol{\mu}_{(1|2)}$) equals

$$\boldsymbol{\mu}_{(1|2)} = \boldsymbol{\mu}_{(1)} + V_{12}V_{22}^{-1}(\underline{\mathbf{x}}_{(2)} - \boldsymbol{\mu}_{(2)}).$$

Note now $\underline{\mathbf{x}}_{(j)}$ denotes the observed values of $\mathbf{x}_{(j)}$.

Proof. Expanding

$$\mathbb{P}^T (V_{11} - V_{12}V_{22}^{-1}V_{21})^{-1} \mathbb{P}$$

with $\mathbb{P} = [\mathbf{x}_{(1)} - \boldsymbol{\mu}_{(1)} - V_{12}V_{22}^{-1}(\underline{\mathbf{x}}_{(2)} - \boldsymbol{\mu}_{(2)})]$ yields

$$(\mathbf{x}_{(1)} - \boldsymbol{\mu}_{(1)})^T (V_{11} - V_{12}V_{22}^{-1}V_{21})^{-1} (\mathbf{x}_{(1)} - \boldsymbol{\mu}_{(1)})$$
$$-(\underline{\mathbf{x}}_{(2)} - \boldsymbol{\mu}_{(2)})^T V_{22}^{-1} V_{21} (V_{11} - V_{12}V_{22}^{-1}V_{21})^{-1} (\mathbf{x}_{(1)} - \boldsymbol{\mu}_{(1)})$$
$$-(\mathbf{x}_{(1)} - \boldsymbol{\mu}_{(1)})^T (V_{11} - V_{12}V_{22}^{-1}V_{21})^{-1} V_{12}V_{22}^{-1} (\underline{\mathbf{x}}_{(2)} - \boldsymbol{\mu}_{(2)})$$
$$+(\underline{\mathbf{x}}_{(2)} - \boldsymbol{\mu}_{(2)})^T V_{22}^{-1} V_{21} (V_{11} - V_{12}V_{22}^{-1}V_{21})^{-1} V_{12}V_{22}^{-1} (\underline{\mathbf{x}}_{(2)} - \boldsymbol{\mu}_{(2)}),$$

which equals the original

$$(\mathbf{x} - \boldsymbol{\mu})^T V^{-1} (\mathbf{x} - \boldsymbol{\mu}) - (\underline{\mathbf{x}}_{(2)} - \boldsymbol{\mu}_{(2)})^T V_{22}^{-1} (\underline{\mathbf{x}}_{(2)} - \boldsymbol{\mu}_{(2)})$$

since

$$V_{21}(V_{11} - V_{12}V_{22}^{-1}V_{21})^{-1} \equiv V_{22}(V_{22} - V_{21}V_{11}^{-1}V_{12})^{-1}V_{21}V_{11}^{-1}$$

(one of the $VA = I$ identities) implies

$$V_{22}^{-1}V_{21}(V_{11} - V_{12}V_{22}^{-1}V_{21})^{-1}V_{12}V_{22}^{-1}$$
$$= (V_{22} - V_{21}V_{11}^{-1}V_{12})^{-1}V_{21}V_{11}^{-1}V_{12}V_{22}^{-1}$$
$$= (V_{22} - V_{21}V_{11}^{-1}V_{12})^{-1}(V_{21}V_{11}^{-1}V_{12} - V_{22} + V_{22})V_{22}^{-1}$$
$$= (V_{22} - V_{21}V_{11}^{-1}V_{12})^{-1} - V_{22}^{-1}.$$

□

Corollary 11.1. *From what we have shown so far it automatically follows that*

$$\det(V) \div \det(V_{22}) = \det(V_{11} - V_{12}V_{22}^{-1}V_{21}).$$

Proof: To demonstrate this explicitly, take the determinant of each side of

$$\begin{bmatrix} I & -V_{12}V_{22}^{-1} \\ O & V_{22}^{-1} \end{bmatrix} \begin{bmatrix} V_{11} & V_{12} \\ V_{21} & V_{22} \end{bmatrix} = \begin{bmatrix} V_{11} - V_{12}V_{22}^{-1}V_{21} & O \\ V_{22}^{-1}V_{21} & I \end{bmatrix}.$$

Example 11.10. Using the normal distribution of the previous Example 11.9, find the conditional distribution of X_1, X_2, and X_3 given that $X_4 = -1.05$ and $X_5 = -5.8$.

Solution.

> evalm(evalf(SubMatrix(V, 1..3, 1..3)) − SubMatrix(V, 1..3, 4..5)
    .MatrixInverse(SubMatrix(V, 4..5, 4..5)).SubMatrix(V, 5..4, 1..3))

$$\begin{bmatrix} 2.4023 & 4.4943 & 2.0690 \\ 4.4943 & 9.2644 & 4.8276 \\ 2.0690 & 4.8276 & 6.0690 \end{bmatrix}$$

> evalm([−1, 2, 0] + SubMatrix(V, 1..3, 4..5)
    .MatrixInverse(SubMatrix(V, 4..5, 4..5))
    .([−1.05, −5.83] − [3, −2]));

$$\begin{bmatrix} -4.6609 & 3.1623 & 1.5924 \end{bmatrix}$$

□

EXERCISES

Exercise 11.1. Consider the following autoregressive model (after equilibration):

$$X_n = 0.9\,X_{n-1} - 0.6\,X_{n-2} + 0.3\,X_{n-3} + \varepsilon_n,$$

where $\varepsilon_n \in \mathcal{N}(0, 13)$. Find:

(a) The first five (up to and including ρ_5) serial correlation coefficients;
(b) The corresponding power spectrum;
(c) $\mathrm{Var}(X_n)$;
(d) The value of the following partial correlation coefficient:

$$\rho(X_n, X_{n-3} \mid X_{n-1}).$$

Exercise 11.2. Let X_1, X_2, X_3, and X_4 have a multivariate normal distribution with respective means of $3.5, -4.5, 0.5$, and 5.5 and a variance–covariance matrix of

$$
\mathbb{V} =
\begin{bmatrix}
34 & 2 & -12 & 29 \\
2 & 40 & -32 & -16 \\
-12 & -32 & 32 & 4 \\
29 & -16 & 4 & 33
\end{bmatrix}
$$

(verify it is positive definite). Find the corresponding correlation matrix.

Exercise 11.3. (Continuation of Question 11.2). What is the conditional distribution of

(a) X_1, X_2, X_4 given that $X_3 = -2.5$;
(b) X_2, X_4 given that $X_1 = 1.5$ and $X_3 = -2.5$;
(c) X_2 given that $X_1 = 1.5$, $X_3 = -2.5$ and $X_4 = 3.25$?

Exercise 11.4. (Continuation of Question 11.2). Find a 4×4 matrix $\mathbb{B}$ such that $\mathbb{B}\mathbb{B}^{\mathrm{T}} = \mathbb{V}$. Generate a random independent sample of 50 quadruplets from the multivariate normal distribution of the previous question.

Exercise 11.5. Consider the Yule model with $\sigma = 3.8$ and the following set of α:

1. $\alpha_1 = 0, \alpha_2 = 0.9$;
2. $\alpha_1 = 1.8, \alpha_2 = -0.9$;
3. $\alpha_1 = -1.8, \alpha_2 = -0.9$.

For each of these:

(a) Determine whether the resulting process is stationary.
(b) Plot its correlogram (of all ρ_k that are "visibly" nonzero).

(c) Generate and plot a sample of 200 consecutive observations (let the process stabilize first).

(d) Use these samples to estimate ρ_1 and ρ_2 and, consequently, α_1 and α_2.

Exercise 11.6 (Continuation of Exercise 11.5). Study the following AR(3) model: $\sigma = 0.17$, $\alpha_1 = 2.8$, $\alpha_2 = -2.705$, $\alpha_3 = 0.9$ (this time, include the estimates of ρ_3 and α_3 as well).

Exercise 11.7. Consider a Yule model with $\alpha_1 = 1.3$, $\alpha_2 = -0.35$, and $\sigma = 0.67$. Find:

(a) $\Pr(X_n > 1.2 \mid X_{n-1} = -0.3 \cap X_{n-2} = -1.2)$;
(b) $\Pr(X_{n+2} > 1.2 \mid X_{n-1} = -0.3)$.

Exercise 11.8. Consider the following autoregressive model:

$$X_n = 0.9\,X_{n-1} - 0.6\,X_{n-2} + 0.3\,X_{n-3} + \varepsilon_n,$$

where $\varepsilon_n \in \mathcal{N}(0, 13)$. Compute:

(a) $\Pr(X_{116} < 10 \mid X_{115} = 7.6 \cap X_{114} = -1.2 \cap X_{113} = 3.1)$;
(b) $\Pr(X_{116} < 10 \mid X_{115} = 7.6 \cap X_{113} = -1.2 \cap X_{112} = 3.1)$.

CHAPTER 12
Basic Probability Review

This chapter is for review purposes only and may be skipped. Those who require an introduction to MAPLE programming should read Chap. 13 first.

12.1 PROBABILITY

A SAMPLE SPACE Ω is the *set of* all possible (complete) *outcomes* (called SIMPLE EVENTS) of a specific random experiment.

An EVENT is a *subset* of the sample space.

A UNION of two events $A \cup B$ is the collection of simple events that belong to A, B, or *both*.

An INTERSECTION of two events $A \cap B$ is the collection of simple events that belong to both A *and* B (sometimes we call it the overlap of A and B).

A COMPLEMENT of an event $\overline{A}$ is the collection of simple events that do *not* belong to A.

An empty subset is called the NULL EVENT, or $\varnothing$.

BOOLEAN ALGEBRA

Unions, intersections, and complements obey the following rules:

1. Both unions and intersections are COMMUTATIVE

$$A \cap B = B \cap A,$$
$$A \cup B = B \cup A$$

and ASSOCIATIVE

J. Vrbik and P. Vrbik, *Informal Introduction to Stochastic Processes with Maple*, Universitext, DOI 10.1007/978-1-4614-4057-4_12, © Springer Science+Business Media, LLC 2013

$$(A \cap B) \cap C = A \cap (B \cap C) \equiv A \cap B \cap C,$$
$$(A \cup B) \cup C = A \cup (B \cup C) \equiv A \cup B \cup C,$$

meaning we do not need parentheses for a union (respectively intersection) of *any* number of events.

2. A union can be DISTRIBUTED over an intersection:

$$(A \cap B) \cup C = (A \cup C) \cap (B \cup C)$$

and vice versa

$$(A \cup B) \cap C = (A \cap C) \cup (B \cap C).$$

Both of these can be generalized, for example,

$$(A \cap B) \cup (C \cap D) \cup (E \cap F \cap G) = (A \cup C \cup E) \cap \cdots,$$

which results in a total of $2 \times 2 \times 3 = 12$ terms.

3. DeMorgan laws:

$$\overline{A \cup B} = \overline{A} \cap \overline{B},$$
$$\overline{A \cap B} = \overline{A} \cup \overline{B}$$

(each of which can be generalized to any number of events).

4. And a few nameless rules:

$$A \cap A = A,$$
$$A \cup A = A,$$
$$A \cap \overline{A} = \emptyset,$$
$$A \cup \overline{A} = \Omega,$$
$$\overline{\overline{A}} = A.$$

PROBABILITY

The probability of a simple event is its *relative frequency* of occurrence in a *long run* of independent replicates of the corresponding random experiment.

The probability of an event $\Pr(A)$ is the sum of probabilities of the simple events that constitute A.

A few rules:

$$\Pr(\overline{A}) = 1 - \Pr(A),$$
$$\Pr(A \cap \overline{B}) = \Pr(A) - \Pr(A \cap B),$$
$$\Pr(A \cup B) = \Pr(A) + \Pr(B) - \Pr(A \cap B).$$

The last of these can be generalized to three or more events; thus:

$$Pr(A \cup B \cup C) = Pr(A) + Pr(B) + Pr(C) - Pr(A \cap B)$$
$$- Pr(A \cap C) - Pr(B \cap C) + Pr(A \cap B \cap C) + \cdots .$$

MUTUAL INDEPENDENCE OF EVENTS

Two events are independent when

$$Pr(A \cap B) = Pr(A) \cdot Pr(B).$$

Three events are independent when any two of them are independent and

$$Pr(A \cap B \cap C) = Pr(A) \cdot Pr(B) \cdot Pr(C).$$

In general, k events are MUTUALLY INDEPENDENT when the probability of any such intersection (of any number of them) equals the product of the corresponding individual probabilities.

CONDITIONAL PROBABILITY

The conditional probability of A, given the actual outcome is inside B, is defined by

$$Pr(A \mid B) = \frac{Pr(A \cap B)}{Pr(B)}.$$

Note $Pr(A \mid B) = Pr(A)$ when A and B are independent.

Often, these conditional probabilities are the natural probabilities of a (multistage) random experiment, a fact utilized by the following PRODUCT RULE:

$$Pr(A \cap B) = Pr(A) \cdot Pr(B \mid A)$$

(the previous formula in reverse). This can be generalized to three or more events:

$$Pr(A \cap B \cap C) = Pr(A) \cdot Pr(B \mid A) \cdot Pr(C \mid A \cap B)$$

$$\vdots$$

A PARTITION of a sample space is a collection of events (say A_1, A_2, ..., A_k) that do not overlap (any two of them have a null intersection) and whose union covers the whole sample space. For any such partition, and any *other* event B, we have the following FORMULA OF TOTAL PROBABILITY:

$$Pr(B) = Pr(A_1) Pr(B \mid A_1) + Pr(A_2) Pr(B \mid A_2) + \cdots + Pr(A_k) Pr(B \mid A_k).$$

RANDOM VARIABLE

A random variable assigns, to each simple event, a *number* (e.g., the total number of dots when rolling two dice). Its DISTRIBUTION is a table listing all possible *values* of the random variable, with the corresponding probabilities. Alternatively, we may *compute* these probabilities via a specific formula, called a PROBABILITY FUNCTION, defined by

$$f_X(i) = \Pr(X = i).$$

This is possible only when the random variable is of a DISCRETE type (the set of its values is either finite or countable – usually consisting of integers only).

When a random variable can have any real value (from a specific interval), it is of a CONTINUOUS type (the individual probabilities are all equal to zero). In that case, we must switch to using the so-called PROBABILITY DENSITY FUNCTION (PDF), defined by

$$f_X(x) = \lim_{\varepsilon \to 0} \frac{\Pr(x \le X < x + \varepsilon)}{\varepsilon}.$$

For a discrete-type random variable X, the total-probability formula reads

$$\Pr(B) = \sum_{\forall i} \Pr(B \mid X = i) \cdot f_X(i)$$

as the set of events $\{X = i, \forall i\}$ constitutes a partition.

The formula can be extended to a continuous-type random variable X thus:

$$\Pr(B) = \int_{\forall x} \Pr(B \mid X = x) \cdot f_X(x) \, dx.$$

MULTIVARIATE DISTRIBUTION

Based on the same random experiment, we can define two (or more, in general) random variables; let us call them X and Y. In the discrete case, their JOINT PROBABILITY FUNCTION is

$$f_{XY}(i, j) = \Pr(X = i \cap Y = j).$$

In the continuous case, this must be replaced by the joint PDF, defined by

$$f_{XY}(x, y) = \lim_{\varepsilon \to 0} \frac{\Pr(x \le X < x + \varepsilon \cap y \le Y < y + \varepsilon)}{\varepsilon^2}.$$

Marginal Distribution

A MARGINAL DISTRIBUTION is a distribution of X, ignoring Y (or vice versa), established by

$$f_X(i) = \sum_{\forall j|i} \Pr(X = i \cap Y = j)$$

in the discrete case and by

$$f_X(x) = \int_{\forall y|x} f_{XY}(x, y)$$

when X and Y are continuous. Note the summation (integration) is over the CONDITIONAL range of Y given a value of X.

Conditional Distribution

The CONDITIONAL DISTRIBUTION of X given that Y has been observed to have a specific value is given by

$$f_X(i \mid Y = \mathbf{j}) = \frac{\Pr(X = i \cap Y = \mathbf{j})}{f_Y(\mathbf{j})}$$

or

$$f_X(x \mid Y = \mathbf{y}) = \frac{f_{XY}(x, \mathbf{y})}{f_Y(\mathbf{y})}$$

in the discrete and continuous cases, respectively. Note the resulting ranges (of the i and x values) are *conditional*. The bold face indicates $\mathbf{y}$ is a fixed (observed) value – no longer a variable.

Example 12.1. Consider the following joint PDF:

$$f(x, y) = 2x(x - y) \quad \text{for} \quad \begin{cases} 0 < x < 1 \\ -x < y < x \end{cases}$$

(zero otherwise).

Find the two marginals. Also, find the conditional distribution of X, given $Y = -\frac{1}{2}$.

Solution.

$$> f_{XY} := \begin{cases} 2 \cdot x \cdot (x - y) & 0 < x < 1 \text{ and } -x < y < x \\ 0 & otherwise \end{cases} :$$

$$> plot3d\,(f_{XY}, x = 0..1, y = -x..x, axes = boxed);$$

PROBABILITY DENSITY FUNCTION

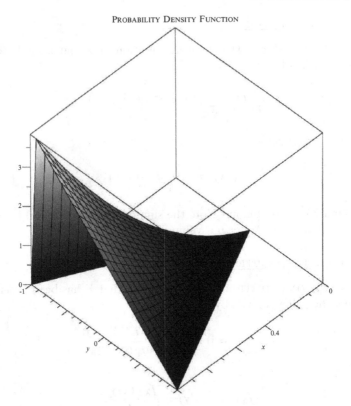

> $f_X := \int_{-\infty}^{\infty} f_{XY} \, dy;$

$$f_X := \begin{cases} 0 & x \le 0, \\ 4x^3 & x < 1, \\ 0 & 1 \le x; \end{cases}$$

> $f_Y := \int_{-\infty}^{\infty} f_{XY} \, dx;$

$$f_Y := \begin{cases} 0 & y < -1, \\ \frac{2}{3} + \frac{2}{3}y^3 - y(1 - y^2) & y \le 0, \\ \frac{2}{3} - \frac{2}{3}y^3 - y(1 - y^2) & y \le 1, \\ 0 & 1 < y; \end{cases}$$

> $plot\,(f_Y, y = -1..1);$

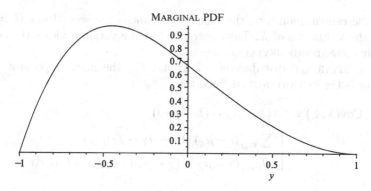

MARGINAL PDF

$$> f_{X|Y=-\frac{1}{2}} := simplify\left(\left.\frac{f_{XY}}{f_Y}\right|_{y=-\frac{1}{2}}\right);$$

{Read $X|Y$ as X given Y.}

$$\begin{cases} 0 & y < -1 \\ \frac{2}{3} + \frac{2}{3}y^3 - y(1-y^2) & y \le 0, \\ \frac{2}{3} - \frac{2}{3}y^3 - y(1-y^2) & y \le 1, \\ 0 & 1 < y; \end{cases}$$

$$> \int_{\frac{1}{2}}^{1} f_{X|Y=-\frac{1}{2}} \, dx;$$

{just to verify the basic property of any distribution, including conditional}

$$1$$

□

MOMENTS

MOMENTS are of two basic types, SIMPLE, defined by

$$\mathbb{E}\left(X^k\right) = \begin{cases} \sum_{\forall i} i^k \cdot f_X(i), \\ \int_{\forall x} x^k \cdot f_X(x) \, dx \end{cases}$$

(discrete and continuous cases, respectively), or CENTRAL,

$$\mathbb{E}\left((X - \mu)^k\right) = \begin{cases} \sum_{\forall i} (i - \mu)^k \cdot f_X(i), \\ \int_{\forall x} (x - \mu)^k \cdot f_X(x) \, dx, \end{cases}$$

where μ is the first ($k = 1$) simple moment, called the MEAN.

Of the central moments, the most important is the second one ($k = 2$), called the VARIANCE of X. The square root of the variance yields the corresponding STANDARD DEVIATION σ_X.

For a BIVARIATE distribution (of X and Y), the most important *joint* moment is the COVARIANCE, defined by

$$\text{Cov}(X, Y) \equiv \mathbb{E}\left((X - \mu_X) \cdot (Y - \mu_Y)\right)$$

$$= \begin{cases} \sum_{\forall i,j} (i - \mu_X) \cdot (j - \mu_Y) \cdot f_X(i, j) \\ \iint_{\forall x,y} (x - \mu_X) \cdot (y - \mu_Y) \cdot f_{XY}(x, y) \, dx \, dy \end{cases}.$$

The corresponding CORRELATION COEFFICIENT is

$$\rho_{XY} \equiv \frac{\text{Cov}(X, Y)}{\sigma_X \cdot \sigma_Y}$$

whose value is always between -1 and 1.

Note when X and Y are *independent*, their covariance (and thus correlation) are equal to *zero* (but not necessarily the converse: zero correlation does *not* imply independence).

Example 12.2. Using the bivariate distribution of the previous example, compute $\text{Cov}(X, Y)$.

Solution.

$$> \mu_X := \int_0^1 \int_{-x}^x x \cdot f_{XY} \, dy \, dx;$$

$$\mu_X := \frac{4}{5}$$

$$> \mu_Y := \int_0^1 \int_{-x}^x y \cdot f_{XY} \, dy \, dx;$$

$$\mu_Y := -\frac{4}{15}$$

$$> var_X := \int_0^1 \int_{-x}^x (x - \mu_X)^2 \cdot f_{XY} \, dy \, dx;$$

$$var_X := \frac{2}{75}$$

$$> var_Y := \int_0^1 \int_{-x}^x (y - \mu_Y)^2 \cdot f_{XY} \, dy \, dx;$$

$$var_X := \frac{34}{225}$$

$$> cov_{XY} := \int_0^1 \int_{-x}^x (x - \mu_X) \cdot (y - \mu_Y) \cdot f_{XY} \; \mathrm{d}y \; \mathrm{d}x;$$

$$cov_{XY} := -\frac{2}{225}$$

$$> evalf \left(\frac{cov_{XY}}{\sqrt{var_X \cdot var_Y}} \right);$$

{it must be inside the -1 and 1 limits}

$$-0.1400$$

$\square$

In the bivariate (and multivariate) case, we can also define CONDITIONAL MOMENTS, for example,

$$\mathbb{E}\,(X \mid Y = \mathbf{y}) \equiv \begin{cases} \sum_{\forall i | j} i \cdot f_X(i \mid Y = \mathbf{j}), \\ \int_{\forall x | y} x \cdot f_X(x \mid Y = \mathbf{y}) \; \mathrm{d}x, \end{cases}$$

replacing f_X by its conditional counterpart (note the summation and integration ranges are also *conditional*).

IMPORTANT FORMULAS

For a LINEAR COMBINATION of random variables, we get

$$\mathbb{E}\,(aX + bY + c) = a\mu_X + b\mu_Y + c,$$

and

$$\mathrm{Var}(aX + bY + c)$$
$$= a^2 \cdot \mathrm{Var}(X) + b^2 \cdot \mathrm{Var}(Y) + 2ab \cdot \mathrm{Cov}(X, Y),$$
$$\mathrm{Cov}(aX + bY + c, dU + eV + f)$$
$$= ad \cdot \mathrm{Cov}(X, U) + ae \cdot \mathrm{Cov}(X, V) + bd \cdot \mathrm{Cov}(Y, U) + be \cdot \mathrm{Cov}(Y, V).$$

These can be easily extended to a linear combination of more than two random variables.

The total-probability formula can be extended to compute a *simple* moment of Y; thus,

$$\mathbb{E}\left(Y^k\right) = \begin{cases} \sum_{\forall i} \mathbb{E}(Y^k | X = i) \cdot f_X(i), \\ \int_{\forall x} \mathbb{E}(Y^k | X = x) \cdot f_X(x) \; \mathrm{d}x. \end{cases}$$

PROBABILITY-GENERATING FUNCTION

A PGF of a discrete (*integer-valued*) random variable X is defined by

$$P_X(z) \equiv \mathbb{E}\left(z^X\right) = \sum_{\forall i} z^i \cdot f_X(i).$$

We can utilize it to compute FACTORIAL MOMENTS of X, namely,

$$\mathbb{E}\left(X \cdot (X-1) \cdot (X-2) \cdots (X-k+1)\right) = P_X^{(k)}(z=1)$$

[the kth derivative of $P_X(z)$, evaluated at $z = 1$]. The two most important cases are

$$\mu_X = \mathbb{E}(X) = P_X'(z=1)$$

and

$$\begin{aligned}
\text{Var}(X) &= \mathbb{E}\left(X^2\right) - \mu_X^2 \\
&= \mathbb{E}\left(X \cdot (X-1)\right) - \mu_X^2 + \mu_X \\
&= P_X''(z=1) - \mu_X^2 + \mu_X.
\end{aligned}$$

Similarly, by expanding $P_X(z)$ in z, we can recover the individual probabilities of the corresponding distribution; thus,

$$P_X(z) = \Pr(X=0) + \Pr(X=1) \cdot z + \Pr(X=2) \cdot z^2 + \Pr(X=3) \cdot z^3 + \cdots.$$

When X and Y are independent, the PGF of $X + Y$ is the *product* of $P_X(z)$ and $P_Y(z)$.

MOMENT-GENERATING FUNCTION

For a continuous-type random variable, the analogous concept is that of a MOMENT-GENERATING FUNCTION, defined by

$$M_X(t) \equiv \mathbb{E}\left(e^{t \cdot X}\right) = \int_{\forall x} e^{t \cdot x} \cdot f_X(x) \, \mathrm{d}x.$$

This time, the kth derivative of $M_X(t)$, evaluated at $t = 0$ (*not* 1) yields the *simple* moments $\mathbb{E}\left(X^K\right)$. This implies

$$\mu_X = \mathbb{E}(X) = M_X'(t=0)$$

and

$$\text{Var}(X) = \mathbb{E}\left(X^2\right) - \mu_X^2 = M_X''(t=0) - \mu_X^2.$$

For X and Y *independent*, the MGF of $X + Y$ is the product of the individual MGFs of X and Y. Also,

$$M_{aX+b}(t) = e^{b \cdot t} \cdot M_X(a \cdot t).$$

For a bivariate distribution, one can also define the *joint* MGF of X and Y; thus,

$$M_{XY}(t_1, t_2) \equiv \mathbb{E}\left(e^{t_1 \cdot X + t_2 \cdot Y}\right) = \iint_{\forall x, y} e^{t_1 \cdot x + t_2 \cdot y} \cdot f_{XY}(x, y) \; dx \; dy.$$

This can then be used to compute joint simple moments $\mathbb{E}\left(X^k \cdot Y^j\right)$ by differentiating $M_{XY}(t_1, t_2)$ k times with respect to t_1 and j times with respect to t_2 and substituting $t_1 = t_2 = 0$.

To invert an MGF (i.e., to find the corresponding PDF), one needs to find its Fourier transform.

Example 12.3. A random variable's MGF is $(1 - 2t)^{-3}$. Find the corresponding PDF.

Solution.

> *with(inttrans):*

> $CF := (1 - 2 \cdot t \cdot \mathbf{I})^{-3}$; {replace each occurrence of t by $t \cdot \mathbf{I}$ (in MAPLE, $\mathbf{I}$ is a purely imaginary number); this converts the MGF into a characteristic function.}

$$CF := \left(\frac{1}{1 - 2\mathbf{I}t}\right)^3$$

> $f := \dfrac{fourier\,(CD, t, x)}{2 \cdot \pi}$;

$$f := \frac{1}{16} x^2 e^{-\frac{1}{2}x} \text{Heaviside}(x)$$

{Heaviside(x) is a function equal to 1 when the argument is positive, zero otherwise.}

> $\displaystyle\int_0^\infty f \; dx$; {Verifying the total probability.}

$$1$$

$\square$

CONVOLUTION AND COMPOSITION OF TWO DISTRIBUTIONS

When X and Y are independent (of the continuous type), the PDF of $X + Y$ is computed by the so-called CONVOLUTION of the two individual

PDFs; thus,

$$f_{X+Y}(u) = \int_{\forall x} f_X(x) \cdot f_Y(u - x) \, dx.$$

This is a symmetric operation, that is, one must obtain the same answer by

$$\int_{\forall y} f_Y(y) \cdot f_X(u - y) \, dy.$$

Example 12.4. Assuming X_1 and X_2 are independent, each having the PDF $f(x) = 1$ when $0 \le x \le 1$, (zero otherwise), find the PDF of $X_1 + X_2$.

Solution.

$$> f := z \rightarrow \begin{cases} 1 & 0 < z < 1 \\ 0 & otherwise \end{cases} \quad :$$

$$> f_{conv} := \int_0^1 f(x) \cdot f(u - x) \, dx :$$

$$> plot\,(f_{conv}, u = 0..2)\,;$$

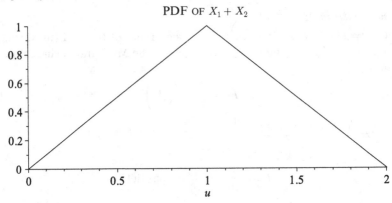

When X_1, X_2, $X_3, \ldots, X_N$ are independent and identically distributed (i.i.d.) random variables and N itself is *random* (of the integer type), the PGF of the sum $S_N = X_1 + X_2 + X_3 + \cdots + X_N$ is

$$P_N\left(P_X(z)\right),$$

assuming the X-distribution is also of the integer type. Otherwise (when the X are continuous), we can find the MGF of S_N to be

$$P_N(M_X(t))$$

This is called the COMPOSITION of the N and X distributions.

Example 12.5. Assuming the X have a binomial distribution with $n = 3$ and $p = 0.3$ and N is Poisson with $\Lambda = 2.6$ (these are reviewed in the following section), plot the distribution of the corresponding S_N.

Solution.

> $P_X := z \rightarrow (.7 + .3 \cdot z)^3$:
> $P_N := z \rightarrow e^{2.6 \cdot (z-1)}$:
> $P_{sum} := P_N(P_X(z))$;

$$P_{sum} := e^{2.6(0.7+0.3z)^3 - 2.6}$$

> $aux := series(P_{sum}, z, 12)$;

$$aux := 0.1812 + 0.2078\,z + 0.2081\,z^2 + 0.1603\,z^3 + 0.1080\,z^4 + 0.0651\,z^5$$

$$+0.0358\,z^6 + 0.0182\,z^7 + 0.0087\,z^8 + 0.0039\,z^9 + 0.0017\,z^{10} + 0.0007\,z^{11}$$

$$+O(z^{12})$$

> $pointplot([seq([i, coeff(aux, z, i)], i = 0..11)])$;

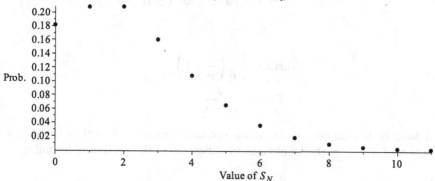

□

12.2 COMMON DISTRIBUTIONS

DISCRETE TYPE

BINOMIAL

X is the number of successes in a sequence of n (fixed) number of BERNOULLI trials (independent, with only two possible outcomes: success, with a probability of p, and failure, with a probability of $q = 1 - p$). We have

$$f(i) = \binom{n}{i} p^i q^{n-i} \quad \text{for } 0 \le i \le n$$

$$\mu = np$$
$$\text{Var}(X) = npq$$
$$P(z) = (q + pz)^n.$$

GEOMETRIC

X is now the number of *trials*, in the same kind of experiment, till (and including) the first success is achieved:

$$f(i) = pq^{i-1} \quad \text{for } 1 \le i,$$

$$\mu = \frac{1}{p},$$

$$\text{Var}(X) = \frac{1}{p}\left(\frac{1}{p} - 1\right),$$

$$P(z) = \frac{pz}{(1 - qz)}.$$

A MODIFIED geometric distribution *excludes* successes; it is thus the distribution of $X - 1$, with the obvious modification of the preceding formulas.

NEGATIVE BINOMIAL

A negative binomial distribution is a distribution of the number of trials needed to achieve k successes:

$$f(i) = \binom{i-1}{k-1} p^k q^{i-k} \quad \text{for } k \le i,$$

$$\mu = \frac{k}{p},$$

$$\text{Var}(X) = \frac{k}{p}\left(\frac{1}{p} - 1\right),$$

$$P(z) = \frac{p^k z^k}{(1 - qz)^k}.$$

A MODIFIED negative binomial distribution is a distribution of $X-k$ (counting *failures* only).

Note a geometric distribution is a special case of a negative binomial distribution, with $k = 1$.

POISSON

A Poisson distribution can be introduced as a limit of binomial distribution, taking $p = \frac{\Lambda}{n}$ and $n \to \infty$:

$$f(i) = \frac{\Lambda^i}{i!} \cdot e^{-\Lambda} \quad \text{for } 0 \leq i,$$

$$\mu = \Lambda,$$

$$\text{Var}(X) = \Lambda,$$

$$P(z) = e^{\Lambda(z-1)}.$$

CONTINUOUS TYPE

UNIFORM

A uniform distribution has a constant probability density in an (a, b) interval; values outside this interval cannot happen:

$$f(x) = \frac{1}{b - a} \quad \text{for } a \leq i \leq b,$$

$$\mu = \frac{a + b}{2},$$

$$\text{Var}(X) = \frac{(b - a)^2}{12},$$

$$\mu(t) = \frac{e^{b \cdot t} - e^{a \cdot t}}{t \cdot (b - a)}.$$

EXPONENTIAL

This is a distribution of X/n where X is geometric, with $p = \frac{1}{n\beta}$ in the $n \to \infty$ limit:

$$f(x) = \frac{1}{\beta} \exp\left(-\frac{x}{\beta}\right) \quad \text{for } 0 \leq x,$$

$$\mu = \beta,$$

$$\mathrm{Var}(X) = \beta^2,$$

$$M(t) = \frac{1}{1 - \beta \cdot t}.$$

Note its memoryless property: the *conditional* distribution of $X - c$, given $X > c$, is exponential with the same mean β as the original X.

GAMMA

A gamma distribution is the distribution of a sum of k independent, exponentially distributed random variables, with the same mean β:

$$f(x) = \frac{x^{k-1}}{\beta^k} \exp\left(-\frac{x}{\beta}\right) \quad \text{for } 0 \leq x,$$

$$\mu = k\beta,$$

$$\mathrm{Var}(X) = k\beta^2,$$

$$M(t) = \frac{1}{(1 - \beta \cdot t)^k}.$$

STANDARDIZED NORMAL

A standardized normal distribution is a distribution of

$$Z = \frac{X_1 + X_2 + X_3 + \cdots + X_n - n \cdot \mu_X}{\sigma_X \cdot \sqrt{n}},$$

where $X_1, X_2, \ldots, X_n$ constitute an i.i.d. sample from any distribution, in the $n \to \infty$ limit (this is called the CENTRAL LIMIT THEOREM):

$$f(z) = \frac{1}{\sqrt{2\pi}} \cdot \exp\left(-\frac{z^2}{2}\right) \quad \text{for } -\infty < z < \infty,$$

$$\mu = 0,$$

$$\mathrm{Var}(X) = 1,$$

$$M(t) = \exp\left(\frac{t^2}{2}\right).$$

GENERAL NORMAL

A general normal distribution can be introduced as a linear transformation of the previous Z; thus,

$$X = \sigma X + \mu.$$

The basic formulas are

$$f(x) = \frac{1}{\sqrt{2\pi} \cdot \sigma} \cdot \exp\left(-\frac{(x-\mu)^2}{2\sigma^2}\right) \quad \text{for } -\infty < x < \infty,$$

$$\mu = \mu,$$

$$\text{Var}(X) = \sigma^2,$$

$$M(t) = \exp\left(\frac{\sigma^2 t^2}{2} + \mu t\right).$$

CHAPTER 13
MAPLE Programming

MAPLE provides an environment for quick evaluation of numeric and symbolic formulas. In a way, MAPLE can be thought of as a calculator that can handle symbols (i.e., unknowns or variables).

We use MAPLE throughout this book to plot, simulate, and carry out the arithmetic of our examples. (The worksheets can be downloaded from extras.springer.com.) We restrict ourselves to as few commands as possible and try to be as literal as the language will allow. For this reason, our MAPLE snippets are almost always suboptimal (in brevity or efficiency) but better for exposition.

The prompt, or the place where one types the input for evaluation, is denoted by ">". Each input line must end with a ";" or ":", the former allowing the result to be printed to the screen (gray and centered), the latter suppressing it. We often represent this input in two dimensions (e.g., $3x^2 + 1$ instead of $3 * x\hat{}2 + 1$). This significantly improves readability but makes the code more difficult to duplicate. For this, we have provided Table 13.1, which shows what to type to obtain each function.

It should be noted MAPLE is extensively documented, and this documentation is easily queried by typing, for example, >? Matrix. We assume the reader will query those commands we do not explicitly introduce.

13.1 WORKING WITH MAPLE

What follows only briefly covers aspects of MAPLE we use. A complete programming guide (for beginners) that is far more comprehensive is available online (free) at MAPLE's Web site: www.maplesoft.com. In addition to guides and documentation, one can also post questions and search the MAPLEPRIMES forums to get expeditious help from the community.

J. Vrbik and P. Vrbik, *Informal Introduction to Stochastic Processes with Maple*, Universitext, DOI 10.1007/978-1-4614-4057-4_13, © Springer Science+Business Media, LLC 2013

Maple Worksheet

A Maple worksheet is a collection of execution lines we assume have been successively processed from top to bottom. These execution lines consist of commands and assignments.

An ASSIGNMENT is of the form

> *name := value :*

Note ":=", not "=", is the symbol for assignment. This is because "=" is the binary equality operator, used in defining equations or doing equality tests.

A fundamental property of (imperative) programming is the ability to recursively reassign named values,

> $a := 2$:

> $a := a + 3$;

$$a := 5;$$

> $a := a * a$;

$$a := 25$$

which we do frequently.

A COMMAND is anything that takes values and returns output (e.g., integration, plotting, arithmetic). We utilize many commands that are labeled in a manner that makes their behavior obvious. Note the output of a command can also be assigned:

> $F := \displaystyle\int x^2 + x + 1 \ \mathrm{d}x$;

$$F := \frac{1}{3} x^3 + \frac{1}{2} x^2 + x$$

What follows is a worksheet.

> $f := x \rightarrow x^2 + 2x + 1$:

> $a := \frac{1}{3}$:

> $f(a)$;

$$\frac{16}{9}$$

Here we have defined a polynomial function f, associated the variable (name) a with the value $\frac{1}{3}$, and then evaluated f at a. As the last line was terminated with ";" MAPLE prints the result (but each line is nonetheless evaluated independently of its printing).

There are some alternatives to using mappings to define functions we often use. The following code illustrates these alternative techniques:

> $f := x^2 + 2x + 1$:

{Now f is just an expression – *not* a mapping/function.}

> $eval\left(f, x = \frac{1}{3}\right)$:
> $f|_{x=\frac{1}{3}}$:
> $subs\left(x = \frac{1}{3}, f\right)$:

are all equivalent.

LIBRARY COMMANDS

Commands to do mathematics/statistics in particular areas are bundled into libraries that must be loaded to be used. It is easy to invoke one (or more) of these libraries:

> $with(Statistics)$;

$[AbsoluteDeviation, AgglomeratedPlot, AreaChart, BarChart, \ldots]$

On calling a package (an alternative name for a library) MAPLE lists all the commands the package makes available. Remember this output can be suppressed by using ":".

Throughout this book we use many commands that are contained in libraries. Since it would be cumbersome to call them in every worksheet, we assume the following packages are loaded at all times (the library names are case sensitive):

1. LinearAlgebra
2. Statistics
3. plots

LISTS AND SEQUENCES

Sometimes we might want to consider a list or sequence of values.

A LIST is an ordering of many values associated with a single name. The individual elements can be retrieved using "[ ]" or by a subscript:

> $A := [x, 2, 3x^2]$:
> A_3;

$$3x^2$$

> $A[3]$;

$$3x^2$$

> A_1;

$$x$$

> A_{-1}; {Negative indices index from the end of the list.}

$$3x^2$$

> $nops(A)$; {List length.}

$$3$$

What is inside the square brackets of a list is a SEQUENCE:

> $B := 1, 2, 3$;

We usually use a sequence when we are trying to build a list:

> $B := NULL$: {Define an empty sequence.}

> $B := B, 1$:

> $B := B, 2$:

> $B := B, 3$;

$$B := 1, 2, 3$$

{and to convert to a list we do}

> $B := [B]$:

Sequences and lists can also be built using the "seq" command. This is particularly useful if you know the closed form (i.e., general pattern) of your list or sequence:

> $c := seq(3 \cdot i, i = 1..4)$;

$$c := 3, 6, 9, 12$$

> $d := \left[seq\left([i, i^2]\right), i = 1..5\right]$;

$$d := [[1, 1], [2, 4], [3, 9], [4, 16], [5, 25]]$$

{Defining a list of lists/ordered pairs is something we do frequently.}

INTEGRAL CALCULUS

A lot of Maple's original design was influenced by the goal of doing symbolic calculus. (Calculus, requiring a lot of tedious symbolic manipulation, was a natural fit within computer algebra systems.)

Unsurprisingly, then, calculus can be done in Maple at the top level (i.e., without calling any libraries). We mostly do derivatives and exact/analytic integrals:

> $f := \frac{1}{2} \cdot x^3 + x^2 + 7$:

> $\int f \, \mathrm{d}x;$

$$\frac{1}{8}x^4 + \frac{1}{3}x^3 + 7x$$

> $\int_1^{10} f \, \mathrm{d}x;$

$$\frac{13167}{8}$$

> $\frac{\mathrm{d}}{\mathrm{d}x} f;$

$$\frac{3}{2}x^2 + 2x$$

We are also able to integrate over an infinite domain or piecewise functions.

PLOTTING

It is informative to visualize functions by way of plots. The simplest approach is to plot a univariate function over a range.

> $f := \dfrac{\sin(x^2) + \cos(x)}{x}:$

> $plot(f, x = -2\pi..2\pi, y = -5..5);$

{When the y-scale is undesirable, we restrict it as well.}

GRAPH OF $f(x)$

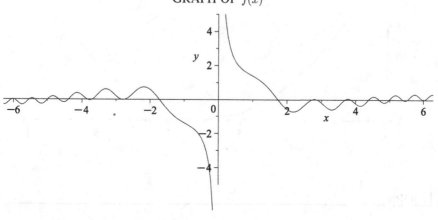

Another way to plot is to provide a list of points.

> $with(plots):$

> $L := \left[seq\left([i, i^2], i = -5..5\right) \right]$:{Parabola, evaluated at integers.}
> $pointplot(L)$;

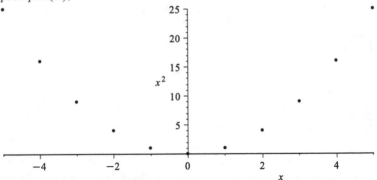

> $L := \left[seq\left(i^2\right), i = -5..5 \right]$:
> $listplot(L)$;

{Here MAPLE assumes you are giving points to be plotted at $[1, L_1]$, $[2, L_2]$, and so on.}

SAME PLOT WITH POINTS CONNECTED.

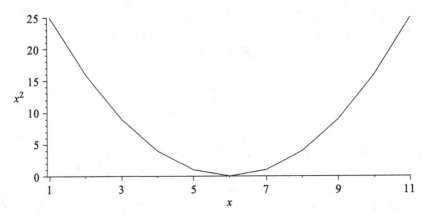

LOOPS

A LOOP is a fundamental construct of computer programming. As its name implies, it allows for a set of commands to be looped or repeated. We use two different types of loops: "for" and "while" loops.

A FOR LOOP is used when you know exactly how many times you want something repeated.

```
> A := [0, 0, 0, 0, 0, 0, 0] :
> for i from  2 to 7 do
    A[i] := i² + A[i − 1];
  end do:
> A;
```

$$[0, 4, 13, 29, 54, 90, 139]$$

A WHILE LOOP is used when you want to loop until a condition is met.

```
> p := 8:
> while not isprime(p) do
    p := p + 1;
  end do:
> p;
```

$$11$$

The while loop can be contingent on several conditions, by using "**and**" and "**or**" to logically tie conditions together.

It is possible to combine these two ideas, that is, to start counting from i until a certain condition is met:

```
> for i from  1 while i ≠ 16 do
    i := 2 · i;
  end do:
> i;
```

$$16$$

This is useful when we want to stop on a condition but also require a counter to keep track of what step we are at.

A few loop tips:

1. Unless you want to see the output for each step of the loop, be sure to close your loop with "**end do:**" not "**end do;**".
2. In the worksheet, to get a new line without executing, do `shift+return`.
3. If you accidentally execute a loop that will never terminate (an infinite loop), then type `ctrl+c` or click the button that looks like a stop sign with a hand in it.

LINEAR ALGEBRA

In MAPLE it easy to work with matrices. The "LinearAlgebra" package offers all the standard functions and transformations one would apply to matrices.

There are two ways to input matrices. The first (preferred) method is to use the matrix contextual menu, which provides an array of clickable cells into which one can enter values. The second method is to provide a list of rows (interpreted to be vectors). Both are demonstrated below.

$$> A := \begin{bmatrix} 2 & 4 & 6 \\ 1 & 3 & 5 \\ 7 & 11 & 13 \end{bmatrix} :$$

$> B := Matrix([[1,2,3],[4,5,6],[7,8,9]]) ;$

$$B := \begin{bmatrix} 1 & 2 & 3 \\ 4 & 5 & 6 \\ 7 & 8 & 9 \end{bmatrix} ;$$

As a matrix is merely a list of lists, we can index its elements by $A[row, column]$ or $A_{row,column}$. We can also extract individual rows.

$> A[1]; \{\text{The first element of } A \text{ is a row.}\}$

$$[2, 4, 6]$$

$> A_{1,2}; \{\text{The second element of } A\text{'s first row/vector.}\}$

$$4$$

Note, using indices, we can also change the values inside a matrix.

```
> for i from 1 to 3 do
    A[i,i] := 0;
  end do:
```

$$\begin{bmatrix} 0 & 4 & 6 \\ 1 & 0 & 5 \\ 7 & 11 & 0 \end{bmatrix}$$

We can also do arithmetic on matrices:

$> A + B; \{\text{Element wise addition}\}$

$$\begin{bmatrix} 3 & 6 & 9 \\ 5 & 8 & 11 \\ 14 & 19 & 22 \end{bmatrix}$$

$> A.B; \{\text{A period computes matrix product.}\}$

$$\begin{bmatrix} 60 & 72 & 84 \\ 48 & 57 & 66 \\ 142 & 173 & 204 \end{bmatrix}$$

> A^4; {Matrix power.}

$$\begin{bmatrix} 18700 & 32588 & 42156 \\ 14696 & 25608 & 33124 \\ 44816 & 78112 & 101060 \end{bmatrix}$$

STATISTICS

We use the "Statistics" package to SAMPLE distributions. This is a two-step process. First we define a RANDOM VARIABLE

> $X := Random\,Variable\,(Normal(0, 1))$:

which enables us to sample the normal distribution by doing

> $Sample(X, 5)$;

$$[-0.216, -0.0558, -0.206, -0.857, -1.049]$$

Of course, we are not restricted to only the normal distribution. In fact, there is a long list of distributions, both discrete and continuous, including "Uniform" and "Exponential", which we use often. Each distribution is specified by one or more parameters and, once converted into a random variable, can be used in an arithmetic expression (which statisticians call a TRANSFORMATION):

> $Sample\left(\dfrac{X}{1 + X^2}, 5\right)$;

$$[0.186, -0.449, -0.379, 0.481, 0.490]$$

We also define our own distributions using "ProbabilityTable". This function takes as input a list, say L, whose values must sum to 1 and returns an integer in $[1, nops(L)]$, taking the elements of L to be the respective probabilities.

For example, the distribution returned by "ProbabilityTable" with $L = \left[\frac{1}{2}, \frac{1}{4}, \frac{1}{4}\right]$ would return 1 with probability $\frac{1}{2}$, 2 with probability $\frac{1}{4}$, and 3 with probability $\frac{1}{4}$.

Usually we only want a single sample value from a distribution. Thus, it is typical that we do

> $Sample(Random\,Variable\,(Normal(0, 1)), 1)_1$;

$$-2.197$$

It is worth mentioning there are other useful library commands like "CDF" and "MGF", which return the cumulative distribution function and moment-generating function, respectively. We avoid using them because it would oversimplify our presentation. However, these functions provide an effective way to verify solutions.

TYPICAL MISTAKES

Symptom/error message	Resolution/explanation
A command is not evaluating; MAPLE just prints what you typed	You have likely misspelled the command or have not invoked the proper library
An equation involving complex numbers or the natural logarithm is not evaluating properly	Be sure to use **I**, **e**, or exp and not *i* and *e* (ordinary names)
`unable to match delimiters`	You have unbalanced parentheses
`invalid subscript selector`	You have tried to access a list position that does not exist
`(in _) unexpected option:`	You have passed too few parameters to a command
`invalid input: _ uses a 1st argument, _ (of type ...`	You have passed too many parameters to a command

Table 13.1: Maple cheat sheet

Input	2D representation	Description
`x*y`	xy	Multiplication
`f/g;`	$\dfrac{f}{g}$	Fractions
`x^ y;`	x^y	Exponents
`x_y;`	x_y	Subscripts
`exp(x);`	e^x	Natural exponent
`f:=x->x^ 2;`	$f := x \rightarrow x^2$	Function definition
`int(f(x),x=a..b);`	$\displaystyle\int_{x=a}^{b} f(x)\,\mathrm{d}x$	Integration
`diff(f(x),x);`	$\dfrac{\mathrm{d}f(x)}{\mathrm{d}x}$	Differentiation
`sum(f(x),i=a..b);`	$\displaystyle\sum_{i=a}^{b} f(i);$	Summation
`mul(f(x),i=a..b);`	$\displaystyle\prod_{i=a}^{b} f(i);$	Product

References

[1] M. S. Bartlett. *An Introduction to Stochastic Processes, with Special Reference to Methods and Applications.* Cambridge University Press, Cambridge/New York, 1980.

[2] U. Narayan Bhat and G. K. Miller. *Elements of Applied Stochastic Processes.* Wiley-Interscience, Hoboken, N.J. 2002.

[3] R. Bronson. *Schaum's Outline of Theory and Problems of Matrix Operations.* McGraw-Hill, New York, 1989.

[4] W. Feller. *An Introduction to Probability Theory and Its Applications.* Wiley, New York, 1968.

[5] S. Goldberg. *Introduction to Difference Equations.* Dover, New York, 1986.

[6] S. Karlin. *A First Course in Stochastic Processes.* Academic, New York, 1975.

[7] S. Karlin and H. M. Taylor. *A Second Course in Stochastic Processes.* Academic, New York, 1981.

[8] J. G. Kemeny and J. L. Snell. *Finite Markov Chains.* Springer, New York, 1976.

[9] J. Medhi. *Stochastic Processes.* Wiley, New York, 1994.

[10] J. Medhi. *Stochastic Models in Queueing Theory.* Academic, Amsterdam, 2003.

[11] S. Ross. *Stochastic Processes.* Wiley, New York, 1996.

[12] A. Stuart. *Kendall's Advanced Theory of Statistics.* Wiley, Chichester, 1994.

J. Vrbik and P. Vrbik, *Informal Introduction to Stochastic Processes with Maple*, Universitext, DOI 10.1007/978-1-4614-4057-4,
© Springer Science+Business Media, LLC 2013

List of Abbreviations

B&D	Birth and death
CTMC	Continuous-times Markov chain
EMC	Embedded Markov chain
FMC	Finite Markov chain
LGWI	Linear growth with immigration
MGF	Moment-generating function
MLE	Maximum likelihood estimators
PDE	Partial differential equation
PDF	Probability density function
PGF	Probability-generating function
SGF	Sequence-generating function
TPM	Transition probability matrix

J. Vrbik and P. Vrbik, *Informal Introduction to Stochastic Processes with Maple*, Universitext, DOI 10.1007/978-1-4614-4057-4, © Springer Science+Business Media, LLC 2013

Index

J. Vrbik and P. Vrbik, *Informal Introduction to Stochastic Processes
with Maple*, Universitext, DOI 10.1007/978-1-4614-4057-4,
© Springer Science+Business Media, LLC 2013